DISCARD

CELLS
THE BUILDING BLOCKS OF LIFE

Stem Cell Research and Other Cell-Related Controversies

Cells: The Building Blocks of Life

Cells and Human Health
Cell Structure, Processes, and Reproduction
The Evolution of Cells
How Scientists Research Cells
Plant Cells
Stem Cell Research and Other Cell-Related Controversies

Stem Cell Research and Other Cell-Related Controversies

PHILL JONES

Stem Cell Research and Other Cell-Related Controversies

Copyright © 2011 by Infobase Learning

All rights reserved. No part of this book may be reproduced or utilized in any form or by any means, electronic or mechanical, including photocopying, recording, or by any information storage or retrieval systems, without permission in writing from the publisher. For information, contact:

Chelsea House
An imprint of Infobase Learning
132 West 31st Street
New York NY 10001

Library of Congress Cataloging-in-Publication Data
Jones, Phill, 1953-
  Stem cell research and other cell-related controversies / Phill Jones.
    p. cm.— (Cells : the building blocks of life)
  Includes bibliographical references and index.
  ISBN 978-1-61753-005-0 (hardcover)
  1. Stem cells—Popular works. 2. Genetic engineering—Popular works. I. Title. II. Series.
  QH588.S83J66 2011
  616'.02774—dc22   2011006946

Chelsea House books are available at special discounts when purchased in bulk quantities for businesses, associations, institutions, or sales promotions. Please call our Special Sales Department in New York at (212) 967-8800 or (800) 322-8755.

You can find Chelsea House on the World Wide Web at http://www.infobaselearning.com

Text design by Erika K. Arroyo
Cover design by Alicia Post
Composition by EJB Publishing Services
Cover printed by Yurchak Printing, Landisville, Pa.
Book printed and bound by Yurchak Printing, Landisville, Pa.
Date printed: September 2011
Printed in the United States of America

10 9 8 7 6 5 4 3 2 1

3 4859 00344 6928

This book is printed on acid-free paper.

All links and Web addresses were checked and verified to be correct at the time of publication. Because of the dynamic nature of the Web, some addresses and links may have changed since publication and may no longer be valid.

On the cover: Embryonic stem cells magnified by a scanning electron microscope

Contents

● ● ●

1	Two Controversies in Scientific Research	7
2	Stem Cell Science and Controversies	15
3	Genetic Engineering of Plants and Animals	29
4	Genetic Modification of Humans	47
5	Property Rights in Genes and Tissues	58
6	Medical Genetic Testing	69
7	Forensic DNA Analysis	81
8	Evolution and Intelligent Design	91
	Glossary	102
	Bibliography	107
	Further Resources	116
	Picture Credits	118
	Index	119
	About the Author	124

1

Two Controversies in Scientific Research

"Scientific controversies are found throughout the history of science," declared the editors of *Scientific Controversies* (2000). "Many major steps in science, probably all dramatic changes, and most of the fundamental achievements of what we now take as the advancement or progress of scientific knowledge have been controversial and have involved some dispute or another."

Sometimes, a possible meaning of a scientific discovery ignites disputes among the general public. Sometimes, the disputes revolve around a method used to acquire knowledge. Experiments on animals and humans are two scientific strategies, or methods, that generate controversies.

ANIMAL EXPERIMENTATION

In his book *The Vivisection Question* (1901), Dr. Albert Leffingwell told the story of a strange experiment. According to a medical school student, a physiology professor had lectured about the way that an animal's fur prevented heat loss and protected an animal against the cold. To demonstrate the point, the professor packed several rabbits in ice to freeze them. A rabbit that had been painted with varnish to render its fur useless slowly died in front of the class, whereas a rabbit with intact fur was revived. Leffingwell wrote:

> [I]n the present civilization of America, there is nothing so satisfactory in the science teaching of certain institutions as an

experiment with torture, for nothing is cheaper than pain.... I confess I do not understand how such an illustration as this is "essential to the progress of medical science," or practiced "only with a view to the prevention of human or animal suffering." It proved nothing that was unknown before. It demonstrated no facts that any student would have hesitated to believe on the authority of his text-book. The reason why we teach with torture is merely, as I have said before, because nothing else is so cheap. It is experimentation of this kind, so cruel in action, so brutalizing in tendency, so void of utility, that should be abolished by law, and made as criminal throughout the United States, as it is in Great Britain to-day.

Humans have experimented with animals at least since second century Rome. During the late nineteenth century in Great Britain, the public outcry against vivisection, the practice of operating on live animals to gain knowledge, led to the first anti-vivisection law. The Cruelty to Animals Act regulated the use of animals in scientific research. Soon after, the newly-formed American Anti-Vivisection Society advocated regulation of the scientific use of animals in the United States. Over the years, the society has also promoted the elimination of classroom demonstrations with animals.

Despite extensive government oversight for animal welfare in scientific research and the proven benefits of animal experiments, protests against any experimental use of animals continues, and some individuals and organizations advocate the development of alternatives to all animal testing. In addition, emerging technologies, such as alteration of genes and production of clones, have raised new concerns about animal welfare.

HUMAN EXPERIMENTATION

In ancient Greece, healers experimented on prisoners, and in ancient Rome, slaves and criminals were the subjects of experiments. During the nineteenth century, the establishment of consistent, scientific research methods and animal experimentation revealed that humans are the best experimental organism for understanding human biology and for testing medical treatments. French physiologist Claude Bernard argued for a way to advance medicine in his book *Introduction to the Study of Experimental Medicine* (1865). He asserted that careful observations of healthy and diseased people would place medicine on a solid scientific basis. Yet Bernard also emphasized the ethical limits to research with humans. "It is our duty

FIGURE 1.1 Animal activist Acacia Waters, dressed as an astronaut with a monkey mask, protests outside the San Francisco Federal Building in California in August 1996. About two dozen protesters were on hand, urging the U.S. Senate to cancel a $33-million project that sends monkeys into space for testing purposes.

and our right to perform an experiment on man whenever it can save his life, cure him or gain him some personal benefit," Bernard wrote. "The principle of medical and surgical morality, therefore, consists in never performing on man an experiment which might be harmful to him to any extent, even though the result might be highly advantageous to science, i.e., to the health of others."

Seventy years later, the Nazis came to power in Germany in the 1930s and disregarded any idea of "medical and surgical morality." The Nazi regime implemented a program of **eugenics**, an attempt to supposedly improve the human race. The government promoted selective breeding

CHIMP CONTROVERSY

For more than a century, animal experimentation has faced often fierce protests from the general public. One aspect of animal research has drawn particularly heavy criticism: invasive experiments with chimpanzees. "[R]esearch on captive chimpanzees is highly controversial, with opponents citing animal welfare, ecological, ethical, scientific and financial objections," said Andrew Knight in his 2008 article in the journal *AATEX* (Alternatives to Animal Testing and Experimentation.) Writing as the director of Animal Consultants International, a group that supports animal advocacy campaigns, Knight urged a ban on chimpanzee experiments. "Some believe," he wrote, "it is precisely the genetic similarities of chimpanzees to humans which are claimed to make them so useful as experimental models that also confer upon them a similar ability to suffer, and that it is unethical to confine and experiment on them."

The Humane Society of the United States advocates an end to chimp testing and echoes Knight's argument. "Chimpanzees are very social, highly intelligent, and proficient in tool use, problem solving, and numerical skills and can even be taught American Sign Language," the group states on its Web site. "Due to the overwhelming evidence of their intelligence and ability to experience emotions so similar to humans, their suffering under laboratory conditions cannot be refuted." During 2010, the Humane Society promoted the passage of the Great Ape Protection Act, a bill introduced in Congress that would phase out harmful research on chimps in U.S. labs.

among individuals considered to have desirable traits and dealt with people who had "hereditary illnesses" by forcibly sterilizing them or by killing them. German doctors also performed pseudoscientific medical experiments on thousands of concentration camp prisoners without their consent. Some were slowly frozen to death, others were deprived of oxygen in high-altitude experiments, and still others were infected with bacteria, viruses, and parasites and treated with experimental drugs, or wounded to simulate battle injuries that were later intentionally infected. Prisoners became test subjects for poisons and bizarre surgical experiments.

After the close of World War II, some Nazi leaders were put on trial in Nuremberg, Germany, for crimes against humanity. One of these trials, known as the Doctors' Trial, began in December 1946 and ended about eight months later. It revealed atrocities committed in the name of medical science. During the trial, several German doctors argued that no international law or agreement distinguished between legal and illegal experimentation on humans. Dr. Leo Alexander, an American doctor who worked with the prosecution, submitted a memo highlighting points that should define ethical research. The memo became the basis for the Nuremberg Code, which outlines ethical guidelines for human experimentation. The first principle of the Nuremberg Code is that the "voluntary consent of the human subject is absolutely essential," a rule that is now known as **informed consent**. Informed consent means that a

(continues on page 14)

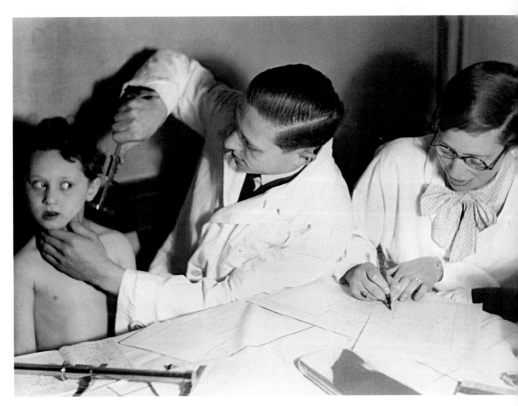

FIGURE 1.2 In an example of eugenics during World War II, a Jewish child is examined by German doctors in Berlin in this 1936 photo.

(*Opposite page*) FIGURE 1.3 A wide variety of animals has contributed to our understanding of human health.

OLD-STYLE HUMAN EXPERIMENTATION REVEALED

In 2010, Susan Reverby, a medical historian at Massachusetts' Wellesley College, discovered the unpublished records of Dr. John C. Cutler. These records revealed experiments that were performed on humans in Guatemala from 1946 to 1948. During this period, American researchers infected hundreds of people with syphilis to test the effectiveness of penicillin. Soldiers, prisoners, mental hospital patients, and prostitutes participated in the research without consent or knowledge of the study's purpose. The researchers must have realized that they performed the study on ethically questionable grounds. "I am a bit, in fact more than a bit, leery of the experiment with the insane people," one of Cutler's supervisors informed him. "They can not give consent, do not know what is going on, and if some goody organization got wind of the work, they would raise a lot of smoke."

News about the 1940s study prompted President Barack Obama to telephone Álvaro Colon, president of Guatemala, and apologize for the actions of U.S. researchers. Hillary Clinton, U.S. secretary of state, and the secretary of health and human services, Kathleen Sebelius, issued a joint public statement. "Although these events occurred more than 64 years ago, we are outraged that such reprehensible research could have occurred under the guise of public health," they said. "The study is a sad reminder that adequate human subject safeguards did not exist a half-century ago."

Such unethical studies were not rare at the time, according to Centers for Disease Control and Prevention director Thomas R. Frieden and National Institutes of Health director Francis S. Collins in *The Journal of American Medical Association* (*JAMA*) commentary. Today, they say, rigorous oversight of research funded or conducted by the U.S. government prevents similar unethical experiments to be carried out. "The 1946–1948 inoculation study should never have happened," they stressed, "and nothing like it should ever happen again."

How Animals Have Helped People

Animal research has contributed to human health in many ways. Among the examples:

Armadillos
Treatment of leprosy

Cats
Methylprednisone for spinal cord injuries

Chimpanzees
Hepatitis B vaccine

Dogs
Discovery of insulin's link to diabetes control
Therapeutic use of antibiotics, such as penicillin, aureomycin, and streptomycin
Development of open heart surgery
Development of the cardiac pacemaker
Coronary bypass surgery
Discovery of antihypertensive drugs
Artificial heart transplantation

Frogs
Chemical communication between cells

Horses
Diphtheria and tetanus vaccine

Monkeys
Discovery of Rh factor
Prevention of poliomyelitis
Rubella vaccine
Development of drugs to control psychophysiological factors in depression, anxiety, and phobias

Pigs
Therapeutic use of cortisone
Beneficial effects of exercise on the heart
Lowering cholesterol

Rabbits
Corneal transplant
Laser treatment to prevent blindness

Rodents
Treatment of rheumatoid arthritis and whooping cough
Development of cancer chemotherapy
Cochlear implants for deafness
Curative drugs for childhood leukemia

Sheep
Development of heart valve replacement

Source: National Institutes of Health Animal Awareness Program
© Infobase Learning

(continued from page 11)

potential subject of an experiment must be educated about the risks of the procedure, must understand the risks so that the individual can make an informed decision, must give consent voluntarily, and must be mentally capable of giving consent.

However, this did not end the controversy. New technologies breed new forms of old controversies. The ability to modify human genes raises the specter of a new strategy for eugenics. While the principle of informed consent seems clear-cut, scientists, however, have now discovered a potential for therapeutic treatments requiring the destruction of human embryos to obtain certain cells. Those who believe that an embryo merits full human rights ask how it can provide informed consent for its own destruction.

2

Stem Cell Science and Controversies

The development of humans and other mammals depends upon the ability of cells to **differentiate** from a generalized cell to a cell with a specialized function. After an egg cell and sperm cell fuse, they form a fertilized egg, or zygote. A zygote is a **totipotent** cell; it can generate all of the types of cells that make an embryo, as well as tissues that provide nutrients to an embryo and support its development. During the first few cell divisions, the zygote produces totipotent cells, but this ability is soon lost. About four days after fertilization, new cells are no longer totipotent, but become **pluripotent** cells. Pluripotent cells can differentiate into any specialized type of cell, except placental cells and other embryo support cells. As pluripotent cells divide, they produce **multipotent** cells, which can differentiate into specialized cells in certain tissues. Neural stem cells, for example, develop into support cells of the nervous system and into nerve cells. Bone marrow contains **hematopoietic stem cells**, which differentiate into cells found in blood. Nerve cells, red blood cells, and hundreds of other specialized cells of the human body are **terminally differentiated**; they do not develop into other types of cells. Therefore, an ability to differentiate into different types of cells decreases during embryonic development:

Totipotent → Pluripotent → Multipotent → Terminally differentiated

The capacity of cells to differentiate into different types of cells is not totally lost in the adult. Muscle, bone marrow, and other tissues have small numbers of multipotent cells that can divide and develop into specialized cells to replace cells that have been injured or worn out.

Stem cells lack the specialized functions of liver cells, kidney cells, and other cells of the body, and yet these cells can develop into specialized cells. Stem cells can differentiate from unspecialized cells into other types of cells by gaining specialized functions and synthesizing specialized proteins, such as proteins of muscle cells that enable contraction. Scientists have had a longstanding interest in stem cells, because the cells can be used to produce replacement tissues and to repair organs of patients with certain diseases. Stem cells have another important characteristic for use in medical treatments: The cells can reproduce themselves, an ability that is sometimes called "self-renewal." Certain types of stem cells can proliferate for more than a year in a laboratory without the cells differentiating. A scientist can obtain a small number of stem cells, and allow the cells to reproduce themselves by cell division in sterile, plastic containers while the cells are immersed in a nutrient-rich liquid. When the **cell culture** contains a suitable number of stem cells, the scientist may coax the cells to differentiate into a certain type, such as a nerve cell.

HUMAN NON-EMBRYONIC STEM CELLS

Researchers investigate two basic types of stem cells, which are named after their source: embryonic stem cells and non-embryonic stem cells. Compared with non-embryonic stem cells, embryonic stem cells usually have a greater capacity to proliferate and remain both pluripotent and undifferentiated for a longer time. Non-embryonic stem cells possess pluripotent stem cell characteristics that vary according to the time of development of the source cells; that is, stem cells obtained from a placenta or umbilical cord more closely possess the qualities of pluripotent stem cells than cells obtained from adult tissue, which are traditionally considered to be multipotent stem cells.

Adult Stem Cells

Scientists have discovered very small numbers of stem cells in the organs and tissues of adults, such as skin, muscle, gastrointestinal tract, retina of the eye, bone marrow, liver, and brain. These **adult stem cells** appear to remain dormant until an injury or a disease damages the tissue. Then, adult stem cells divide, with each stem cell producing two daughter cells. One daughter cell remains a stem cell, continuing a process of self-renewal that ensures the future availability of stem cells. The other daughter cell continues to divide and produces a population of cells that differentiates into the specialized cells of the tissue where they dwell. Hematopoietic adult stem cells of bone marrow, for example, can proliferate and differentiate into white blood cells, platelets, and red blood cells:

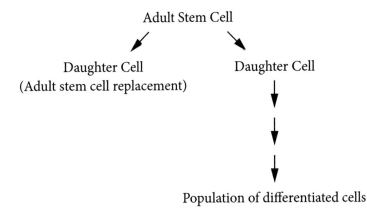

For a long time, researchers considered adult stem cells to be multipotent. However, recent experiments suggest that adult stem cells from one type of tissue can be coaxed into differentiating into cells of another type of tissue. For instance, hematopoietic stem cells can be stimulated to develop into nerve cells, liver cells, or muscle cells. Some scientists propose that certain adult stem cells may be naturally pluripotent. For the present, however, adult stem cells are said to be multipotent and, under certain conditions, may exhibit the property of pluripotent stem cells.

The medical community widely uses treatments that employ types of adult stem cells. For instance, bone marrow stem cells are used to replace blood cells as part of therapy for cancer. One of these therapies treats leukemia, which is a type of cancer in which white blood cells function abnormally and no longer support the immune system's defense of the body by attacking bacteria, parasites, toxins, and viruses. One way to treat leukemia is to use toxic drugs and radiation to destroy a patient's bone marrow, which is producing the abnormal white blood cells. A physician may then administer healthy bone marrow stem cells donated by another person. The bone marrow cells migrate through the patient's bloodstream and to the bones, where they produce white blood cells that function normally.

A patient's immune system can cause problems with a bone marrow transplant. An immune system's white blood cells distinguish between cells that belong in the body and cells foreign to the body. Just as cattle carry the brand of their owner, a body's cells carry identification molecules. These markers can be found on the surface of human cells. A person's immune system attacks a cell that has surface markers that differ from those on the body's own cells. For example, the immune system detects the presence of foreign markers on bacteria, viruses, and parasites. A molecule that rouses the immune system into attack mode is called an **antigen**. Certain white blood cells react to an antigen by attacking cells that have the antigen on their surface. Other white blood cells react to an antigen by making antibodies. An **antibody** is a protein that binds to an antigen and directs other white blood cells to attack. Antibodies tag foreign cells for destruction. When human cells from one person are injected into the body of another person, the immune system will probably detect the new cells. If the new human cells lack the proper identification antigens, then the immune system attacks.

Bone marrow is not the only source of white blood cells. The cells are also produced and stored in the spleen and other tissues. A patient's own white blood cells may recognize transplanted bone marrow cells as foreign cells and attack them. Here, the body is said to "reject" the new bone marrow. A doctor can reduce the risk of bone marrow rejection by transplanting tissue with identification markers that match the patient's antigens as closely as possible. For example, a patient's brother, sister, or parent may have suitable bone marrow. A person who is unrelated to the patient may also have bone marrow with suitable antigens. The chance that the patient's body will accept the donor bone marrow increases with the number of matching identification antigens.

A second complication of a bone marrow transplant is caused by the new white blood cells. The transplanted bone marrow can produce white blood cells that identify cells in other body tissues as foreign. The white blood cells can attack the tissues of the patient, which is a complication called graft versus host disease. This is another reason why doctors try to match identification antigens between the donor's tissue and the patient's tissue.

The traditional approach for coping with tissue rejection is to continuously suppress the immune system with drugs that can produce harmful side effects. Physicians and scientists are devising new stem cell therapies to overcome the problem of tissue rejection by using a patient's own stem cells. For instance, a doctor had a patient whose broken ankle would not heal. The doctor removed a sample of bone marrow from the patient's pelvic bone, concentrated the cells, and injected them into the patient's ankle.

BLOOD PHARMING

The Pentagon's research arm, the Defense Advanced Research Projects Agency (DARPA), initiated its Blood Pharming program with the goal of developing new technologies for production of red blood cells in the lab. The program's ultimate aim is to develop a cell culture and packaging system that can use human cells to produce large quantities of universal donor red blood cells (Type O) for transfusing soldiers who are wounded on the battlefield.

As part of its Blood Pharming program, DARPA funded the biotechnology firm Arteriocyte, located in Cleveland, Ohio, which developed a lab-grown blood product. Company scientists started with umbilical cord hematopoietic stem cells, grew them in culture with nutrients, and used a new process to increase the number of cells by 250 times. The scientists then stimulated the cells to differentiate into red blood cells. According to the company, the mass-produced red blood cells are indistinguishable from normal red blood cells. In 2010, Arteriocyte sent a sample of its Type O blood product to the U.S. Food and Drug Administration for evaluation.

Within four months, the bone marrow stem cells had healed the broken ankle. In another example, scientists performed a study with 112 people who had suffered cornea damage to one of their eyes from an accidental chemical burn. The scientists removed stem cells from the healthy eye of each patient and allowed the cells to proliferate in culture dishes. After growing cells in culture, the researchers transplanted each patient's own cells into their damaged eyes. The cells differentiated into corneal tissue and restored sight in 78% of the patients.

Umbilical Cord Stem Cells

At one time, the umbilical cord was disposed as waste tissue after an infant's birth. However, it was discovered that umbilical cord blood is a source of stem cells. Today, **cord stem cells** are routinely collected from the umbilical cord. As multipotent cells, cord stem cells can develop into a limited number of cell types. For example, cord stem cells can differentiate into blood cells, including platelets, which play a vital role in clot formation, white blood cells

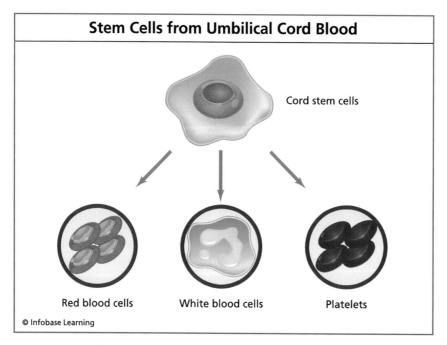

FIGURE 2.1 Umbilical cord blood stem cells can develop into red blood cells, white blood cells, and platelets.

of the body's immune system, and red blood cells, which transport oxygen from lungs to tissues, and waste carbon dioxide from tissues to lungs.

Physicians use cord blood stem cells to treat more than 40 diseases, including leukemia and other cancers, blood disorders, and immune system deficiencies. In the future, scientists may find ways to stimulate cord cells to differentiate into nerve cells for the treatment of Alzheimer's disease and other neurological disorders. Compared with adult stem cells, umbilical cord stem cells are less like to stimulate a rejection response in a patient. The immature cord stem cells may not carry antigens that incite a patient's immune system. Furthermore, white cells produced from cord blood stem cells are less likely to attack a patient's cells. A drawback of umbilical cord blood stem cells is that cord blood contains very few cells, and the stem cells proliferate slowly in the laboratory.

HUMAN EMBRYONIC STEM CELLS

After decades of devising methods for deriving stem cells from mouse embryos, scientists isolated stem cells from human embryos in 1998. In these studies, James A. Thomson, an anatomy professor at the University of Wisconsin, used human embryos that had been created with in vitro fertilization procedures and stored by infertility clinics. The donors, who no longer needed the embryos, donated them for research. U.S. fertility clinics store hundreds of thousands of frozen embryos that are not required for their original purpose.

Following the procedures developed by Thomson and his colleagues, researchers obtain human embryonic stem cells from four or five day-old embryos. At this time in development, a human embryo is a hollow ball of

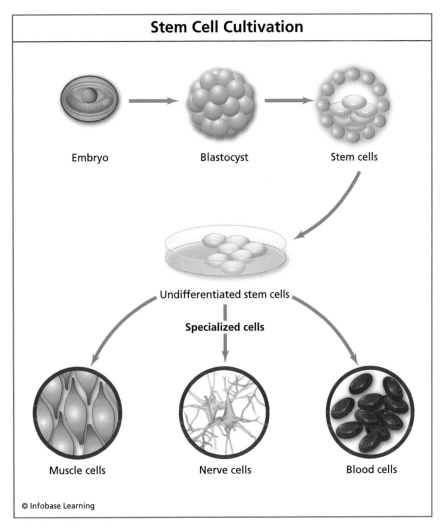

FIGURE 2.2 Embryonic cells can be removed from a blastocyst and grown in a Petri dish in a lab. These undifferentiated stem cells can then be grown to form many types of cells, including neurons.

cells, which is called a blastocyst. A group of about 30 cells, which is called the inner cell mass, is located within the blastocyst cavity. The inner cell mass is removed from the blastocyst and transferred to a sterile plastic culture dish that contains a nutrient liquid. The cells are allowed to divide until they crowd the dish. In a process called subculturing, the cells are gently removed and distributed among new culture dishes. Again, the cells are allowed to proliferate until the populations grow too large for their culture dishes, requiring another division of cells into new dishes. After six months of subculturing, the original group of embryo cells can yield millions of embryonic stem cells that are pluripotent. An **embryonic stem cell line** is a group of embryonic stem cells that proliferates in culture for months or years, remains undifferentiated, and retains pluripotent capacity.

For tissue replacement, human embryonic stem cells offer many advantages, compared with non-embryonic stem cells, such as pluripotent capacity and high rates of proliferation. Potentially, human embryonic stem cells can be allowed to multiply in culture dishes and then stimulated to differentiate into any desired cell type. The new, differentiated cells can then be transplanted into a patient to replace tissue lost from disease or injury. Scientists must overcome many challenges before human embryonic stem cell therapy becomes a standard method of treatment:

- Human embryonic stem cells must be stimulated to differentiate into the correct cell type.
- After transplantation into a patient, the cells must integrate into the correct tissue and perform their functions as part of that tissue.
- Transplanted, differentiated cells must not revert into pluripotent stem cells that proliferate beyond the needs of the tissue. Uncontrolled cell proliferation leads to the development of tumors.
- Ideally, transplanted cells do not carry antigens that stimulate an immune response, leading to a rejection of the cells. This possibility arises, because the stem cells are derived from donated embryos, which would not have the same antigens as those displayed in the patient's cells.

Despite these challenges, the potential of therapies based upon human embryonic stem cells is so significant that scientists continue to work toward overcoming technical barriers. The U.S. Food and Drug Administration sparked hope for this type of therapy in 2010 when the agency

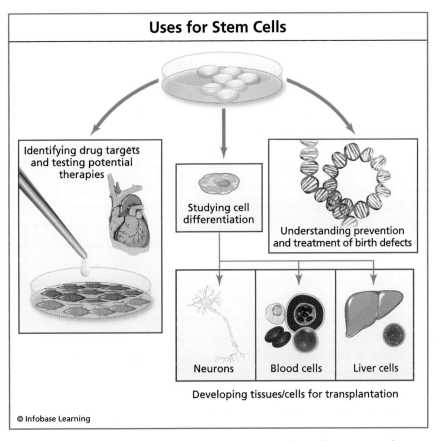

FIGURE 2.3 Proponents of stem cell research note the technique provides a wealth of opportunities to improve human health.

approved the first test in humans of a treatment derived from human embryonic stem cells. Developed by Geron Corporation (located in Menlo Park, California) and the University of California, Irvine, this therapy aims to help patients with recent spinal cord injuries. Scientists working on this project stimulate embryonic stem cells to develop into nerve system cells, which are injected into the spinal cord to repair nerve cells and enable damaged nerves to transmit signals. The goal of the approved study is to show that the therapy is safe for treatment of humans.

CONTROVERSIES ABOUT HUMAN EMBRYONIC STEM CELL RESEARCH

Human embryonic stem cells appear to possess great potential as the basis for a wide variety of therapies to treat human diseases and disorders.

However, debates over the U.S. government's funding of human embryonic stem cell research have slowed progress, at least within the United States.

The roots of the debate can be traced to 1979 when, under pressure from anti-abortion groups, the U.S. Health and Human Services Department disbanded an advisory board that reviewed federally-funded research on human embryos. Presidents Ronald Reagan and George H.W. Bush blocked federal funding for all research on human embryos, a position that reflected their anti-abortionist attitudes. In 1992, President William J. Clinton announced that he would lift the ban on embryo research. He assigned the National Institutes of Health (NIH) the task of drafting guidelines for studies with human embryos. The NIH established the Human Embryo Research Panel, which consisted of 19 scientists, physicians, ethicists, lawyers, and community representatives. The panel developed policies for methods that researchers should use to obtain embryos, and they determined the scope of ethical embryo research. As the panel worked on its policies, Clinton issued an executive order that government-funded scientists would not be allowed to create human embryos for research, which restricted scientists to unneeded embryos stored by in vitro fertilization clinics. During the next several years, the U.S. Congress passed legislation that limited financial support for human embryo research, including a ban on the federal funding of research in which a human embryo is destroyed or discarded, a restriction that eliminated the funding of studies with excess embryos stored at infertility clinics.

Controversy boiled over in 1998 when research groups announced that they had devised methods for culturing human stem cells. The University of Wisconsin's James A. Thomson led one of the research teams. His group obtained early human embryos, which had been donated by couples who had participated in in vitro fertilization programs. Since a private company had funded the research, Thomson and his team avoided the ban on federal funding for human embryo research.

While the journal *Science* hailed human stem cell research as the breakthrough of the year, others condemned the efforts. In its 1999 position statement, The Center for Bioethics and Human Dignity presented their basic argument against human embryonic stem cell research. "Human embryos are not mere biological tissues or clusters of cells," the group asserted. "They are the tiniest of human beings. Thus, we have a moral responsibility not to deliberately harm them." Others promoted a "slippery slope argument": If human embryo research is deemed acceptable, then one day, society may accept human experiments, such as those performed by the Nazis. Many who protested against human embryonic stem cell research pointed to adult stem cells as a more ethical alternative for stem cell-based therapies.

Proponents of human embryonic stem cell research argue that studies require the use of spare, frozen embryos stored at infertility clinics. It is unethical, they say, to ban research with these embryos that could bring about new medical treatments and alleviate human suffering. Pro-life advocate Senator Orrin G. Hatch (R-UT) has said that:

> I believe that human life begins in the womb, not a petri dish or refrigerator. It is inevitable that in the [in vitro fertilization] process, extra embryos are created that will simply not be implanted in a mother's womb. To me, the morality of the situation dictates that these embryos, which are routinely discarded, be used to improve and extend life. The tragedy would be in not using these embryos to save lives when the alternative is that they will be slated for destruction.

Opponents of federal funding for human embryonic stem cell research argue that legislators should not make a moral decision on behalf of all taxpayers to subsidize human embryo experimentation. Why not sidestep the issue by allowing private companies to support such studies? In response, in his 1998 testimony before the U.S. Senate, James Thomson asserted that federally funded research was required to realize new stem cell therapies:

> How soon such therapies will be developed will depend on whether there is public support of research in this area. Private companies will have an important role in bringing new [embryonic stem] cell-related therapeutics to the marketplace; however, the current ban in the U.S. on the use of federal funding for human embryo research discourages the majority of the best U.S. researchers from advancing this promising area of medical research.

Controversies about federal funding persist through the new millennium. In 2000, the NIH announced new guidelines under which the agency would fund research with human pluripotent stem cells derived from embryos created by in vitro fertilization and in excess of clinical need. The NIH would not fund the act of deriving stem cells from human embryos, because it requires the destruction of embryos. On August 9,

2001, President George W. Bush announced his own policy. Federal funds would only be used to support research with existing human embryonic stem cell lines that had been derived from embryos created for fertility treatment and that had been deemed no longer needed. Moreover, couples who had donated the embryos must have given their consent without any payment. Scientists protested that the 22 cell lines that met the criteria would not be sufficient to support significant human stem cell research. Among other problems, the cells were probably infected with viruses and had grown so long in culture that they were beginning to accumulate mutations. (A **mutation** is an alteration in the DNA that can alter proteins encoded by the DNA.) Some of the cell lines had mutations that made it likely that the cells could form tumors if transplanted into a human.

On March 9, 2009, President Barack Obama issued an executive order that revoked the Bush stem cell policy. "[W]hen it comes to stem cell research," Obama said, "rather than furthering discovery, our government has forced what I believe is a false choice between sound science and moral values. In this case, I believe the two are not inconsistent. As a person of faith, I believe we are called to care for each other and work to ease human suffering. I believe we have been given the capacity and will to pursue this research—and the humanity and conscience to do so responsibly." Obama directed the NIH to issue new guidelines for funding human stem cell research. The NIH returned to the old policy of allowing federal funding of human embryonic stem cell research as long as scientists did not use federal funds for the act of obtaining stem cells from human embryos.

Nightlight Christian Adoptions, the Christian Medical Association, and others filed a lawsuit alleging that the new NIH guidelines were illegal. In August 2010, a federal district court judge agreed that one cannot separate the derivation of embryonic stem cells from research with the stem cells, because the derivation is an integral step in performing research. The judge blocked Obama's executive order and the NIH guidelines. Several weeks later, an appeals court lifted the ban until it could decide the matter.

HUMAN EMBRYONIC STEM CELLS OR HUMAN NON-EMBRYONIC STEM CELLS?

Certain therapeutic uses of non-embryonic stem cells, such as bone marrow transplantation, have become standard practices. These medical applications take advantage of multipotent stem cells that differentiate into a restricted number of cell types. However, these non-embryonic stem cells have characteristics that limit their usefulness: The cells can be recovered in only very small numbers, and the cells do not proliferate to a great extent in culture. Stem cell replacement therapies require large numbers of cells. Human embryonic stem cells lack the problems

HOW CAN YOU MEND A BROKEN HEART?

By mid-2010, about 1,000 heart attack patients had been treated with stem cells in clinical trials worldwide. In most of the studies, physicians used stem cells from bone marrow tissue. According to Professor Michael Schneider of Imperial College London, the procedures have been successful with regard to safety, but bone marrow stem cells supplied little improvement in heart function. Schneider leads a research team that investigates heart stem cells, which may provide an improved therapy. Their studies generated a technique to identify rare human heart stem cells, so that the cells could be isolated and purified for transplant into a patient to repair heart damage. These adult stem cells can be stimulated to differentiate into heart muscle cells or cells that build blood vessels. In the next phases of their project, Schneider and his colleagues will extract and purify heart stem cells in the lab, allow the cells to proliferate in culture, and use the cells to repair heart damage.

Other research teams are devising techniques to improve heart treatments with bone marrow stem cells. Dr. Andre Terzic of the Mayo Clinic in Rochester, Minnesota, leads a team that has invented a method for programming human bone marrow stem cells to differentiate into cells that closely resemble heart cells. In their studies, the researchers removed samples of bone marrow stem cells from heart disease patients and identified cells that showed an unusually effective capacity to repair heart tissue. Further examination revealed that the highly regenerative stem cells synthesized excess amounts of certain proteins. The researchers used this information to identify a molecular signature characteristic of the unusually effective stem cells. Next, they treated ordinary stem cells with a combination of proteins and small chemicals to induce the signature within the cells, effectively reprogramming ordinary cells into the super cells. After transplantation into mice, the reprogrammed stem cells repaired and healed heart tissue.

of non-embryonic stem cells: A small number of embryonic stem cells can proliferate indefinitely in cultures. As pluripotent cells, embryonic stem cells can differentiate into virtually any human cell type required for therapy. Although technical hurdles bar an immediate application of therapies using human embryonic stem cells, scientists are overcoming these barriers. Human embryonic stem cell therapy poses another type of challenge, one that science cannot resolve: The use of stem cells derived from human embryos incites opposition based upon religious beliefs and moral convictions.

The debate about human embryonic stem cell research revolves around the question about the moral status of an early embryo. Some argue that an early embryo has the full status of a human being. The Roman Catholic Church, for example, considers life as beginning at conception. Others insist that an early embryo is a collection of undifferentiated cells and lacks moral status. For instance, Jewish, Islamic, Hindu, and Buddhist traditions hold that the moral status as a human being occurs much later in the development process. Many scientists, physicians, and ethicists take a compromise position that an early embryo is a *potential* human being. Although an early embryo should not be treated as a person, it should be handled with dignity.

It is unlikely that the debate will be resolved in the near future. Several U.S. states have even passed legislation banning human embryonic stem cell research. Views differ across the globe, as well. Some countries have laws that allow research with stem cells derived from unused in vitro fertilization embryos, and some countries forbid it. Some countries allow the creation of human embryos for research.

3

Genetic Engineering of Plants and Animals

A BRIEF OVERVIEW OF GENE EXPRESSION

Cells store information in the **nucleic acid** molecule, called **deoxyribonucleic acid (DNA)**. Bacteria typically have a single, circular DNA molecule that contains the cell's genetic material. This DNA molecule is the organism's **chromosome**. Plant and animal cells have linear chromosomes arranged in pairs within a membrane-bound structure called the **nucleus**. The number of chromosomes contained within a nucleus varies among species. A fruit fly has four pairs of chromosomes, a rice plant has 12 pairs, and human cells have 23 pairs.

A DNA molecule is a **polymer**, which is a large chemical formed by combining smaller chemical units. The chemical units of a polymer are linked with each other by **covalent bonds**. A covalent bond is a strong form of chemical bond in which two atoms share electrons. In the case of DNA, covalent bonds link **nucleotides**. Each nucleotide has three parts: (1) a **deoxyribose** sugar molecule, which is a five-carbon sugar molecule called **ribose** that is missing a particular oxygen atom; (2) a phosphate molecule, a chemical group that contains phosphorus; and (3) a molecule called a **base**, which contains nitrogen.

The sugar group of one nucleotide binds with the phosphate group of another nucleotide to form a "sugar-phosphate-sugar-phosphate" structure, which is called the sugar-phosphate backbone of DNA. The bases of nucleotides stick out from the sugar-phosphate backbone. A DNA molecule has four types of bases: adenine, cytosine, guanine, and thymine, which

30 STEM CELL RESEARCH AND OTHER CELL-RELATED CONTROVERSIES

are represented by the first letter of their names: ACGT. The sequence "AGCTGA," for example, indicates a small piece of DNA that has the base sequence "adenine-guanine-cytosine-thymine-guanine-adenine."

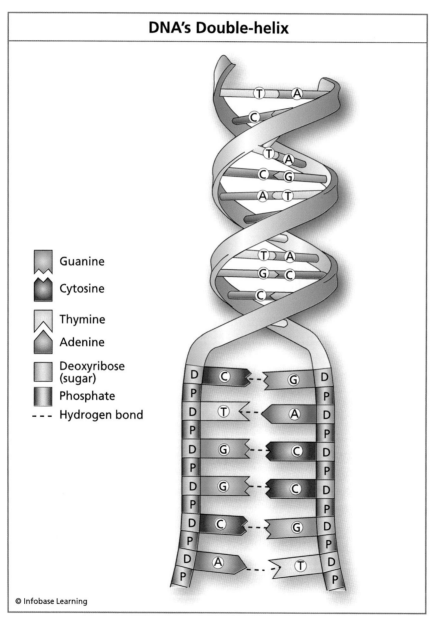

FIGURE 3.1 The structure of DNA is shown here. Note the base pairs of nucleotides that make up the "rungs" of the ladder.

Two strands of DNA can form a double-stranded helix, because certain bases are attracted to each other in a manner that can be imagined as a type of magnetic attraction. The rules of attraction are simple: An A on one DNA strand pairs with a T on the other DNA strand, and a G on one DNA strand pairs with a C on the other DNA strand. When bases of two different DNA strands bind together, they form a **base pair**. Consider a very short DNA molecule with two strands. One strand has the following sequence: CATTAGCATGGACT. The other strand would have the sequence GTAATCGTACCTGA. Together, the strands would appear as follows:

CATTAGCATGGACT

GTAATCGTACCTGA

This is the case because the first C in CATTAGCATGGACT pairs with the first G in GTAATCGTACCTGA, the first A in CATTAGCATGGACT pairs with the first T in GTAATCGTACCTGA, and so on. The base pairs, AT and GC, have the same overall shape. Since they have the same shape, an AT base pair and a GC base pair can fit into any order between the two sugar-phosphate backbones without deforming the helix.

A **gene** is a DNA nucleotide sequence that provides the information a cell needs to synthesize a **protein** or a **ribonucleic acid (RNA)** molecule. Scientists can alter plants and animals by manipulating **gene expression**, which is the process in which information stored in DNA is used to synthesize proteins and RNA molecules that affect the function or structure of a cell.

Like DNA, RNA is a polymer of nucleotides. However, RNA and DNA differ in three ways. First, RNA has a base called uracil that takes the place of thymine in DNA. For example, the sequence AGA TGT CCT in a piece of DNA would appear as AGA UGU CCU in an RNA molecule. A second difference between DNA and RNA is that DNA contains deoxyribose sugars, whereas RNA contains ribose sugars. A third difference is that RNA usually exists in the form of a single strand, whereas DNA can be found as a double-stranded helix.

To produce a protein, the data stored in DNA must be transferred to the cell's protein production machinery. Two stages of **protein synthesis** are **transcription** and **translation**. During transcription, data is transferred from DNA by the synthesis of RNA molecules called **messenger RNA (mRNA)**. Messenger RNA has a nucleotide sequence that is a copy of a nucleotide sequence found in a DNA molecule. (An mRNA molecule will not carry an exact copy of a DNA molecule's nucleotide sequence, because DNA uses thymine, whereas RNA uses uracil.) Messenger RNA molecules carry their recipes for a protein in the form of a **genetic code**. The genetic code uses the four nucleotide bases found in RNA, which are organized into triplets of bases called **codons**. Triplets of four nucleotide bases provide 64 combinations (4 x 4 x 4). A protein is a polymer of **amino acids**. Since cells typically use 20 types of amino acids to synthesize a protein, 64 types of codons are more than sufficient to encode proteins. Some amino acids are encoded by two or more codons. For example, the amino acid leucine is encoded by the codons UUA, UUG, CUU, CUA, CUC, and CUG. Not all codons stand for an amino acid; some codons act as stop signals for protein synthesis. During the second stage of protein synthesis, translation, a cell's protein synthesis machinery translates codons in an mRNA molecule to produce a sequence of amino acids in a protein polymer.

Genetic Engineering

The word *engineering* refers to the use of science to design and construct things. Genetic engineering uses science to isolate, analyze, and modify genes. Genetic engineering relies upon **recombinant DNA** technology. Recombinant DNA is DNA that has been altered in the laboratory by the addition or deletion of nucleotide sequences. Scientists who work in this field transfer foreign genes into various organisms. The transferred gene is often called a **transgene**, while an organism that receives a transgene may be called a **transgenic** organism, or a genetically engineered organism.

One vital tool in the genetic engineer's toolbox is a collection of **restriction enzymes**. An **enzyme** is a protein that increases the rate of a chemical reaction. A restriction enzyme is an enzyme that cuts DNA at a specific (restricted) place. A restriction enzyme glides along the backbone of a DNA molecule until it comes across a certain target nucleotide sequence that is called its **cleavage site**. There, the enzyme binds to the DNA molecule. Once the enzyme has the DNA backbone in its firm grasp, the enzyme twists into a different shape. As the enzyme contorts, it distorts the DNA molecule and breaks the DNA backbone. Different restriction enzymes bind to different cleavage sites. By using a collection of various restriction enzymes, a genetic engineer can cleave DNA at selected places.

Scientists often convey a transgene into a cell using a DNA molecule called a **vector**. One type of vector is a **plasmid**, which is a small, circular DNA molecule that replicates itself in bacterial cells. During the early 1970s, Stanley Cohen, a scientist at Stanford University in California, and Herbert Boyer of the University of California, San Francisco, published their classic paper about a plasmid experiment that established recombinant DNA technology. In one study, they inserted frog genes that encode certain RNA molecules into a plasmid and inserted the modified plasmid into *Escherichia coli* bacteria. Then, they isolated RNA from the *E. coli* bacteria and found that the bacteria had produced RNA from the frog genes, proving that the frog transgenes functioned in transgenic bacteria.

The researchers prepared their plasmid vector using the restriction enzyme *Eco*RI. This enzyme seeks out the nucleotide sequence GAATTC. *Eco*RI breaks a DNA molecule after the guanine (G) nucleotide in the cleavage site. In a double-stranded DNA molecule, the cleavage site would appear as follows:

.... GAATTC ....

.... CTTAAG ....

In the Cohen-Boyer study, *Eco*RI cut open a circular plasmid. Cleavage of the DNA left two short stubs of single-stranded DNA, called "sticky ends." Sticky ends can form base pairs with themselves or with matching pieces of DNA.

.... G AATTC ....

.... CTTAA G ....

*Eco*RI then cut the frog genes at both ends, creating two sticky ends.

AATTC .... G

G .... CTTAA

After mixing the cleaved plasmid and frog genes, the two types of DNA fit together like pieces of a puzzle by forming base pairs at the sticky ends.

.... GAATTC .... GAATTC ....

.... CTTAAG .... CTTAAG ....

An enzyme called ligase sealed the breaks in the DNA molecules. Restriction enzymes are like scissors and ligase enzymes are like glue.

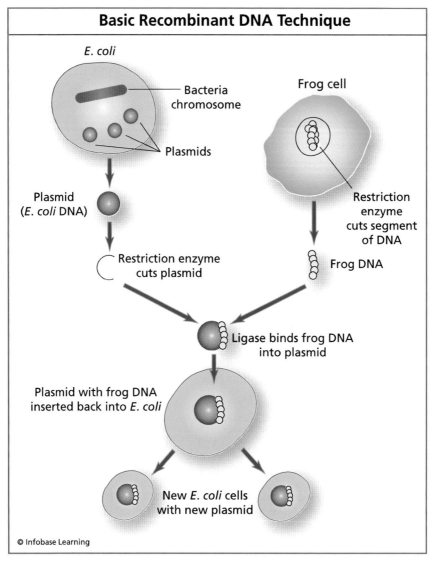

FIGURE 3.2 In a groundbreaking gene-splicing experiment, Stanley Cohen and Herbert Boyer broke up cells of a common bacterium, E. coli, and took out small, ring-shaped pieces of DNA called plasmids. They then used a restriction enzyme to cut open the plasmids. They used the same enzyme to produce segments of DNA from the cells of frogs. The bacterial and frog DNA segments joined together because of the complementary "sticky ends" of single-stranded DNA attached to each segment. Boyer and Cohen used a ligase, another type of enzyme, to bind the segments together, creating a new plasmid that contained frog as well as bacterial DNA. The researchers then inserted the plasmids carrying the foreign genes into other E. coli bacteria and showed that the bacteria produced RNA molecules from the foreign genes. When the bacteria multiplied, the added genes were duplicated along with the bacteria's own genetic material.

Genetic Engineering of Plants and Animals 35

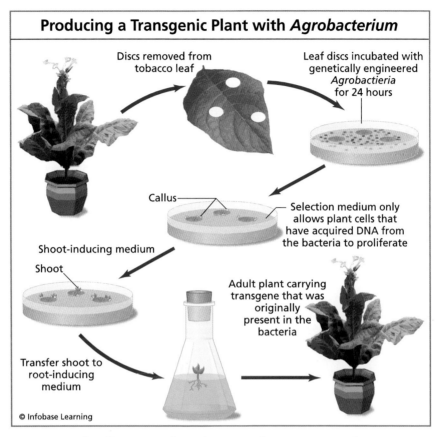

FIGURE 3.3 This illustration shows how to make a transgenic plant.

The Cohen-Boyer experiments formed the foundation of the basic genetic engineering process:

- Cleave a gene from a DNA molecule with restriction enzymes
- Splice the gene into a plasmid or other vector
- Insert the vector into a host cell to produce a transgenic cell
- Culture transgenic cells that produce copies of the vector

Scientists have applied this process to alter plants, especially plants grown for food.

GENETIC ENGINEERING OF CROPS

During the early 1980s, scientists devised a way to use a parasite to produce transgenic plants. A soil-dwelling bacteria, *Agrobacterium tumefaciens*, infects plant cells by inserting a plasmid into the plant's chromosomes. To produce transgenic plants, scientists modified the *Agrobacterium* plasmid into a vector that can be used to insert any gene into a plant cell. Since most plant cells are totipotent, a complete plant can be grown from one plant cell. After treating plant cells with an *Agrobacterium* vector, scientists culture the cells in plastic dishes in a laboratory. The cells are used to make a transgenic plant. If a transgenic plant produces viable seed, the transgene will be passed on to new generations.

Another popular technique for producing a transgenic plant is microprojectile bombardment, or biolistics, a method in which DNA is shot into a plant cell. The method is performed with DNA-coated gold particles that have diameters of about 0.000032 inches (0.8 microns). The particles are loaded into a gene gun, and high pressure helium gas propels the DNA-coated pellets into plant cells at speeds up to 2,000 feet per second (600

FIGURE 3.4 An unidentified protester holds a sign reading "No Franken-Food!," as busloads of Biotechnology Industry Organization conference attendees arrive at an opening night reception in San Francisco on June 6, 2004. Protesters, who object to the proliferation of genetically modified food, often plan large demonstrations during the biotechnology industry's annual conference.

meters per second). After the DNA-coated particles burst through a plant cell wall, DNA molecules are released from the particles and become inserted within plant cell chromosomes.

In 1996, farmers planted the first significant large-scale transgenic crop. Ten years later, farmers in 22 countries planted transgenic crops that covered over 252 million acres (101.9 million hectares). Most of the crops were transgenic soybeans, corn, cotton, canola, and alfalfa. Farmers in the United States grew a little over half of the world's transgenic crops. During 2010, the U.S. Department of Agriculture reported that the majority of corn (86%), cotton (93%), and soybeans (93%) planted by U.S. farmers that year had been genetically altered.

Many of the first transgenic crops produce proteins that kill insect pests. This eliminates the need for farmers to spray crops with traditional chemical insecticides, which are not only costly for farmers but also harmful to the environment. For example, certain genetically engineered crops synthesize toxic proteins produced by the soil-dwelling bacterium, *Bacillus thuringiensis*. These toxic proteins typically kill a limited number of insect species and do not directly affect other animals. Genetic engineers have also designed transgenic plants that can resist viruses that cause diseases. Other transgenic plants have been designed for increased flavor or nutritional content. For instance, Golden Rice is transgenic rice that contains beta-carotene in the grain. Since the human body converts beta-carotene to vitamin A, Golden Rice is beneficial for people who have a vitamin A deficiency.

Some transgenic crops have been engineered to synthesize proteins and chemicals for medicine and industry. These genetic engineering efforts are often called molecular farming, or biopharming, and can provide a means to produce therapeutic proteins and chemicals at lower cost and in greater amounts than traditional methods.

The production of transgenic crops has incited longstanding protests. Those who oppose transgenic crops highlight possible risks to human health and the environment. One argument against transgenic crops is that altering genes might accidentally enable a plant to produce a molecule toxic to humans. Another argument is that cultivating transgenic plants that make insecticidal proteins could harm the environment by killing insects that do not feed on crops. The death of these non-target insects would affect birds, fish, and other animals that eat the insects.

Yet another basis for protesting transgenic crops is that the plants may spread outside of their fields. In 2010, scientists presented the first

FRANKENFOOD

In 2004, California's Mendocino County became the first county in the United States to forbid the presence of genetically engineered crops and animals. The county declared itself a genetically modified organism-free zone. Other U.S. counties and cities considered joining the GMO-free movement by banning "Frankenfood."

Why do so many people oppose genetically engineered food? Carol Tucker Foreman suggested several reasons in her article published in the March 2004 issue of *The American Enterprise*. Foreman, who was the director of the Food Policy Institute at the Consumer Federation of America, noted that the agricultural biotechnology industry and government regulators argue that "biotech food" products are based on solid science, "are good for farmers, safe for consumers, and often beneficial for the environment." These arguments are unpersuasive, Foreman wrote, because they ignore the powerful cultural and personal attachments that most people have to their food. "From the apple in the Garden of Eden to the Golden Arches, food—what, how, and in what quantity we eat—has played a central role in our lives. We eat to live but we also live to eat. Food is more than fuel for the body. Since what we eat literally becomes part of our bodies, food is the source of some of our greatest pleasures and, not surprisingly, greatest fears."

Detractors of engineered crops often promote the idea that scientists who apply genetic modification tread in realms best left to the Creator. As Prince Charles, the Prince of Wales, wrote in the June 8, 1998, edition of *The Daily Telegraph*:

> The fundamental difference between traditional and genetically modified plant breeding is that, in the latter, genetic material from one species of plant, bacteria, virus, animal or fish is literally inserted into another species, with which they could never naturally breed. The use of these techniques raises, it seems to me, crucial ethical and practical considerations.
>
> I happen to believe that this kind of genetic modification takes mankind into realms that belong to God, and to God alone.

evidence that transgenic canola plants established populations outside of their original fields. They also found two cases in which canola plants contained multiple transgenes, which suggests that different types of genetically engineered plants can breed in the wild to produce new types of transgenic plants. The study supports protestors who have claimed that farmers cannot contain transgenic crops and their transgenes.

ALTERATION OF ANIMALS BY GENETIC ENGINEERING AND CLONING

Genetic Engineering of Animals

In December 1982, the journal *Nature* included an article by scientists Richard Palmiter, Ralph Brinster and their colleagues that reported the production of a transgenic mouse. The researchers had injected a modified rat growth hormone gene into fertilized mouse eggs and then transferred the eggs into a mouse that acted as a foster mother. Although all of the pups were of normal birth size, some of them rapidly grew to nearly twice the size of their litter mates under the influence of the growth hormone transgene. Since then, scientists have engineered many types of transgenic mice, as well as transgenic insects, fish, goats, sheep, chickens, cows, rabbits, and pigs.

The Palmiter-Brinster team used the technique of DNA microinjection to produce their transgenic mice. The technique is performed by allowing a sperm cell to fuse with an egg cell. For a brief time, the fused cell contains two pronuclei: one egg nucleus and one sperm cell nucleus. Then, the pronuclei combine to create one nucleus with a full set of chromosomes. Before the two pronuclei fuse, a scientist injects a piece of DNA into one of the pronuclei. The DNA fragment, which includes one or more transgenes, inserts itself into the chromosomes. The pronuclei fuse and the cell continues to develop as an embryo. This microinjection technique remains a popular method for producing transgenic rats and mice.

Scientists also produce transgenic animals with a vector made from a virus. A normal virus carries the instructions in its genetic material to reproduce itself, but lacks the means to do so. Instead, a virus must infect a cell and take over the cell's protein and nucleic acid synthesis machinery to make copies of itself. To construct a vector, scientists alter a virus's genetic material by removing genes for virus reproduction. They next add one or more transgenes. A virus vector can efficiently deliver transgenes to the chromosomes of young animal embryos.

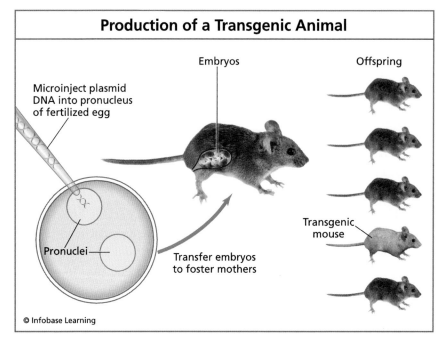

FIGURE 3.5 To produce a transgenic animal, DNA is microinjected into a pronucleus of a fertilized egg. The microinjected eggs are then transferred to foster mothers and allowed to develop. Some of the offspring are transgenic, in that they have incorporated the injected DNA into their genome.

Researchers engineer transgenic animals for a variety of purposes:

- *Animal models*: While some transgenic mice are designed for research into basic questions of biology, others provide animal models for the study of the causes of human diseases and to test possible treatments. Transgenic pigs have been engineered to study diseases, such as cystic fibrosis, and to test methods for treating damage to arteries caused by high cholesterol blood levels.
- *Livestock improvements*: Researchers have produced transgenic cows that synthesize protein-enriched milk, and transgenic pigs that provide more healthful types of meat. Researchers at the University of Guelph (Ontario, Canada) developed a pig that synthesizes a bacterial enzyme, which allows the animals to absorb more phosphorus from feed. The transgenic pigs excrete less phosphorus and decrease a pollution problem created by

First, she would obtain a mouse egg cell and remove the nucleus from the cell. Then, the scientist would take a skin cell from Nicky, isolate the nucleus from the skin cell, and transplant it into the egg cell. The altered egg cell contains Nicky's entire supply of nuclear DNA. Next, the scientist would transfer the egg cell to a foster mother mouse. Theoretically, the mother should give birth to a young version of Nicky. The reality of cloning, however, is much more complicated.

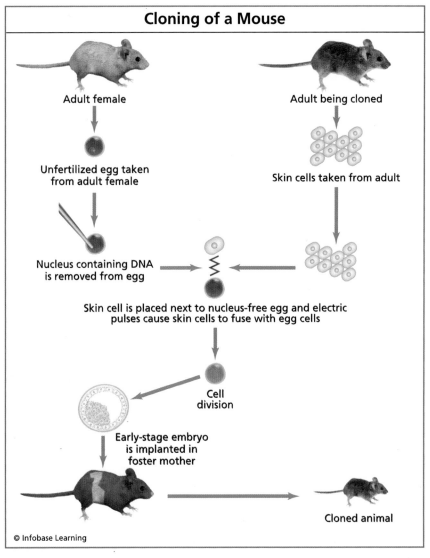

FIGURE 3.6 The steps to cloning a mouse are outlined, including manipulation of the adult female's egg and implantation in a foster mother.

conventional pigs. Other research efforts aim to produce transgenic livestock that resist diseases, such as mad cow disease.
- **Production of pharmaceuticals**: Some transgenic pigs, goats, sheep, and cattle secrete protein drugs into their milk. The therapeutic proteins can be isolated from the milk, purified, and used as medicines for humans.
- **Improved xenotransplants**: Progress in organ transplantation has led to a severe shortage of organs available for transplant. According to the U.S. Organ Procurement and Transplantation Network, more than 100,000 people in the United States alone waited for organ transplants during 2010. Xenotransplantation—the transplant of tissue from a nonhuman—can provide one solution to the shortage of human organs. However, the human immune system poses a significant difficulty with xenotransplants. The immune system attacks cells that are foreign to the body. Scientists have engineered transgenic pigs with human proteins on their cells that should prevent an immune attack after pig tissue is transplanted into humans.
- **Pest control**: Researchers are engineering insects in a variety of ways, such as the alteration of cotton-plant-eating pink bollworms to carry a gene that would prevent their offspring from developing, modification of bacteria in the kissing bug's gut to kill the parasite that causes Chagas' disease, and the alteration of mosquitoes to render the insects incapable of transmitting malaria.

Cloning Animals

In 1997, scientists at Scotland's Roslin Institute announced the birth of a cloned sheep named Dolly. In this context, a **clone** is an animal that is a copy of another animal. Dolly, cloned with genetic material from a mature sheep, was the first mammal to be cloned from an adult animal. *Time* magazine reported the news in an article entitled "The Age of Cloning." (Dolly died in February 2003 from tumors caused by a virus.) However, Dolly's cloning failed to signal the start of a time when animal cloning became common. An age of cloning has yet to appear.

The idea of the cloning technique is simple: Swap one nucleus for another nucleus. After this **nuclear transfer**, the cell with the new nucleus should develop according to instructions in the new genetic material. For example, suppose a scientist wants to clone her pet mouse, named Nicky.

While the cloning technique is simple, the outcome is unpredictable. The success rate of cloning can be very low. Hundreds of attempts may be required before a researcher obtains a living clone. Even when the technique produces a clone, the animal may not be identical to the animal that donated the nucleus. Identical nucleotide sequences are insufficient to guarantee success in cloning.

Two cells with identical DNA can produce different proteins due to **epigenetic** effects. An epigenetic effect is a change in nuclear DNA that affects the activity of genes without changing nucleotide sequences in DNA. One type of epigenetic change is DNA methylation. A methyl group is a cluster of one carbon atom and three hydrogen atoms. In DNA methylation, an enzyme attaches a methyl group to a cytosine base in a DNA molecule. Attachment of methyl groups can interfere with the cell's machinery that synthesizes RNA from DNA. By interfering with gene transcription, methylation stops the production of the protein encoded by the gene.

During the early 1980s, scientists discovered that epigenetic changes not only affect gene expression during the life of a cell, but can also affect gene expression of an offspring's cells. This is a phenomenon named **genomic imprinting**: For a small number of genes, the parental source of a gene affects gene activity. Scientists chose the term "imprinting" to signal that something inactivates a gene without changing the nucleotide sequence of the gene. An imprinted gene is inactive in the sense that transcription from the gene is turned off. Two genes encoding the same protein can be active or inactive, depending upon whether it is inherited from the father or from the mother. That is, some genes have an imprint in the set of genes inherited from the mother, while other genes have an imprint in the set of genes inherited from the father. DNA methylation and other epigenetic alterations play a role in genomic imprinting.

Epigenetic changes can profoundly affect the outcome of cloning. North Carolina State University researchers, for example, discovered that low birth weight and long-term harmful health effects in cloned pigs are linked to at least three imprinted genes.

CONCERNS ABOUT ENGINEERED AND CLONED ANIMALS

The genetic engineering of animals has fueled protests for decades. Many claim that it is unethical to alter animals with recombinant DNA technology. Scientists argue that humans have altered animals by selective breeding for thousands of years. As one example, humans have bred a variety of dogs, resulting in radical changes of anatomy—just compare a Chihuahua with its ancestor, the wolf.

However, genetic engineering differs from traditional breeding methods in several significant ways. Unlike traditional methods that require generations of selective breeding, the new technology can produce striking alterations in a single generation. Another difference between the two approaches is that in traditional breeding, the selected trait may be accompanied by a number of genes that support the desired trait. With genetic engineering, a desired trait may require a single transgene. The introduction of this single, new gene may disrupt the normal balance of traits within the transgenic animal. For example, scientists have developed transgenic salmon that have an extra growth hormone gene and grow twice as fast as conventional (non-transgenic) salmon. Sometimes, rapidly growing transgenic salmon develop severe deformities, such as enlarged disfigurements of the skull. This type of unintended effect found in transgenic animals has been used to support objections to genetic engineering based on animal welfare concerns.

The engineering of farm animals has also stirred anxiety about food safety. A transgenic animal may synthesize new proteins, which may be included in food produced from these animals. Will these new proteins cause an allergic reaction in people who consume the food? The U.S. Food and Drug Administration (FDA) monitors food safety, including the possibility that food from transgenic animals may cause an allergic reaction.

The FDA has created its own controversies in how it regulates food from engineered animals. For years, advocates of a consumer's "right to know" have protested the agency's refusal to establish compulsory labeling of food products from transgenic animals. The FDA's policy is that labeling is not required if food from a transgenic animal does not differ chemically from food obtained from a non-transgenic animal.

Transgenic animals present environmental issues. If transgenic animals escape into the wild, they could spread their transgenes by mating with non-transgenic animals of the same species. By the time that an unintended spread of transgenes is detected, it may be too late to remedy the situation. Transgenic animals might replace their non-transgenic counterparts.

Animal cloning has provoked concerns about animal welfare. Cloning efforts have been plagued by low success rates and the production of cloned animals with increased susceptibility to infections, high rates of tumor growth, heart disorders, and other health problems. Food safety has been another issue raised by animal cloning. The FDA has stated that the agency does not have food safety concerns about products from cattle,

MITOCHONDRIA: A CELL'S POWERHOUSE AND A CSI TOOL

Plant and animal cells divide tasks required for survival among membrane-bound, organ-like structures called **organelles**. The nucleus, which contains the majority of a cell's DNA, is one type of cell organelle. **Mitochondria** are another type of organelle. These jelly bean-shaped structures process the molecules they obtain from food to supply energy for the cell. A typical human cell contains one nucleus and hundreds of identical mitochondria. Each mitochondrion has four or five copies of a circular, 16,500-base pair molecule of DNA that includes about 37 genes. Mitochondrial genes encode enzymes required for the organelle's function.

Mammals have different patterns of inheritance for nuclear DNA and mitochondrial DNA. During fertilization, a sperm's nuclear DNA enters the egg and combines with the egg's nuclear DNA so that the offspring inherit DNA from the mother and father. In contrast, mitochondrial genes are usually inherited only from the mother. Most of a sperm's mitochondria are left outside the egg at the time of fertilization. Although small numbers of sperm mitochondria enter the egg, paternal mitochondrial DNA is usually destroyed. Consequently, the fertilized egg typically contains only maternal mitochondrial DNA.

Analysis of mitochondrial DNA offers a unique contribution to criminal investigation. Biological samples that lack a sufficient amount of nuclear DNA for analysis may still contain enough mitochondrial DNA. Consequently, scientists can analyze mitochondrial DNA from bones and teeth subjected to high temperatures and other harsh conditions. Criminal investigators often rely upon an analysis of mitochondrial DNA recovered from stored evidence of old, unsolved cases ("cold cases"). The technique does have a limitation: Unlike nuclear DNA, mitochondrial DNA is not unique to an individual. Typically, maternally inherited mitochondrial DNA is identical among maternally related people, such as a brother and a sister, or a mother and her child.

swine, or goat clones, or their offspring. Nevertheless, protests against the marketing of such food continue.

WHAT NEXT?

Using recombinant DNA technology, scientists can engineer transgenic plants and animals with new traits. While these efforts aim to benefit mankind, many people protest against genetic engineering. The engineering of transgenic animals has particularly provoked dissent, perhaps, because the technology fuels apprehension about future applications. For instance, can technology for engineering animals be used to produce transgenic humans?

4

Genetic Modification of Humans

A person's **genome**—the collection of genes in chromosomes found in the nuclei of a person's cells—determines the color of their eyes, hair, skin, and other physical traits. Genes also affect the risk of developing a disease. A **genetic disease** results from DNA mutations that create abnormal nucleotide sequences in a person's genome. Genetic mutations can lead to disease in two general ways. Sometimes, a mutation in a gene directly causes illness by affecting a protein that is vital for health. Other genetic diseases arise from a combination of factors: A DNA mutation that places a person at risk for a disease may be followed by exposure to something in the environment that promotes the development of a disorder. Disease-causing environmental factors include infection by certain viruses, exposure to certain toxic chemicals, and overexposure to the sun's ultraviolet light.

Four types of diseases are caused entirely or partly by a change in DNA: single-gene disorders, multigene disorders, chromosomal disorders, and mitochondrial disorders. Scientists have identified more than 10,000 human single-gene diseases, which are caused by a mutation in one gene. In these disorders, the protein encoded by the mutated gene may be synthesized in an altered form, or the protein may not be synthesized at all. Examples of single-gene disorders include sickle cell anemia, galactosemia, and cystic fibrosis. Multigene disorders are caused by mutations in two or more genes and may involve exposure to environmental factors. High blood pressure, arthritis, cancer, and diabetes are examples of multigene disorders. Chromosomal disorders result from changes in chromosome structure. Chromosome alterations include a change in the order of genes and the duplication or lack of an entire chromosome, or a large part of a chromosome. Down's syndrome and Cri-du-Chat syndrome are two examples of chromosomal disorders.

Whereas the other genetic diseases result from changes in nuclear DNA, mitochondrial disorders arise from mutations in the small mitochondrial DNA molecule. Scientists have identified more than 250 mitochondrial DNA mutations associated with a disease. Mitochondria produce chemical energy for a cell's many activities. A mutation that hinders this function particularly affects muscle cells and nerve cells, which require great amounts of energy. Consequently, a mitochondrial disease often leads to hearing and vision loss, seizures, and muscle weakness.

GENE THERAPY

Scientists apply genetic engineering technology to devise cures, or if a cure is not possible, treatments for genetic diseases. For more than 20 years, scientists and physicians have been testing gene therapies to treat single-gene diseases. The objective of **gene therapy** is to correct the altered genes that are responsible for the development of disease. Four possible strategies may be used to correct a mutated gene:

- Insert a transgene into the patient's genome, which encodes a protein that the patient's body does not synthesize. After treatment, the patient's cells should produce the missing protein.
- Regulate gene expression of a patient's mutated gene that produces a mutated protein. For example, inactivating a mutated gene would help a patient if that gene encoded an altered form of a protein that harmed the patient.
- Repair a mutated gene.
- Swap a mutated gene with a copy of a normal, healthy version of the gene.

Many scientists have tackled the first strategy: inserting a transgene that will synthesize a protein that is missing in a patient. To perform this type of gene therapy, a scientist must develop a delivery system to carry a transgene to the patient's cells. A popular type of delivery system uses genetically engineered viruses. In this technique, a scientist typically modifies the genetic material of a virus in at least two ways: (1) delete nucleotide sequences that contain instructions for making copies of the virus, and (2) add nucleotide sequences that encode the transgene. The genetically engineered virus then delivers the transgene to the cells. Once inside the cells, the virus's nucleic acid molecule inserts into the cell's chromosomes, and the transgene enables cells to produce the therapeutic protein.

Physicians use two tactics for administering a transgene into a patient. In one approach, called **ex vivo** gene therapy, a physician obtains human cells, such as stem cells or differentiated cells, which may be obtained from the patient. The cells are treated with DNA that contains a transgene to produce transgenic cells. The physician then administers the transgenic cells to the patient. The other approach, which is called **in vivo** gene therapy, is performed by administering a transgene directly to a patient.

In 1990, physicians performed the first gene therapy trial for the inherited disease known as ADA deficiency. This disorder is caused by a

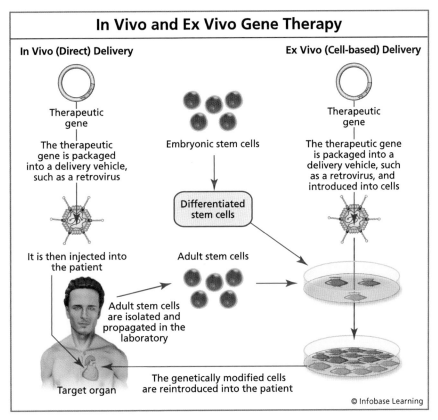

FIGURE 4.1 Gene therapy can treat a person who has an abnormal gene response for disease development.

mutation that impairs the function of the enzyme adenosine deaminase (ADA). Normally, ADA degrades the toxic chemical, deoxyadenosine. When a person lacks functional ADA enzyme, toxic chemicals accumulate in the blood and kill white blood cells, which the body relies upon to fight infections of viruses and bacteria. A person who has ADA deficiency suffers repeated and severe infections, which can prove fatal. In one case, doctors performed the gene therapy trial by removing white blood cells from a four-year-old girl who had ADA deficiency. They treated her cells with genetically engineered virus DNA that contained a normal ADA gene, so that the transgene inserted into the chromosomes of the white blood cells. Then, the doctors infused the transgenic white blood cells into their patient. They repeated this treatment about a dozen times over the next several years. Eventually, the patient's body produced about 25% of the normal amounts of ADA enzyme. More recently, physicians have introduced a normal ADA gene into hematopoietic stem cells. After an infusion of transgenic cells into the patient, the stem cells divide and produce blood cells that synthesize ADA enzyme.

ADA gene therapy exemplifies the ex vivo gene therapy approach to treating a genetic disease. In 2007, physicians used in vivo gene therapy to treat a vision disorder caused by a mutation in the RPE65 gene, which disrupts the function of cells in the eye's retina. To treat the disease, doctors inserted a needle through patients' eyes and into their retinas. Then, they injected virus vectors that contained copies of the normal RPE65 gene. Later studies showed that the treatment can improve eyesight, especially for children who have inherited the retinal disease.

Following the 1990 gene therapy trial for ADA deficiency, physicians have applied gene therapy technology to test treatments for a variety of disorders. According to the *Journal of Gene Medicine*, more than 1,000 gene therapy clinical trials were approved worldwide from 2000 to 2010. Most of these therapies were aimed at the treatment of cancer and other multigene disorders.

Experience with gene therapy revealed that treatment can yield unpredictable results, which has raised concerns about safety. For example, researchers have designed many gene therapy vectors with altered virus nucleic acid. When these viruses deliver a transgene to human DNA, the transgene inserts in a random place in a chromosome. The risk is that a random insertion can activate a gene that harms the patient. For example, two children who took part in a gene therapy trial in France appear to have developed cancer because viral DNA had inserted in a way that promoted the growth of cancer cells. Epigenetic changes also play a role in the success of gene therapy. For example, DNA methylation can inactivate a therapeutic transgene.

GENETIC MODIFICATION OF HUMAN GERMLINES

In theory, a genetic disease may be treated by two types of gene therapy. The methods discussed above are sometimes called somatic gene therapy because the technique introduces a transgene into **somatic cells** (body cells). Some have raised a concern that somatic gene therapy could *accidentally* result in the transfer of transgenes into germ cells—egg cells or sperm cells. If this occurred, then the transgene would be passed on to future generations. A technique called **germline genetic modification** is the *deliberate* introduction of a transgene into germ cells. The goal of this approach would be to cure a genetic disease in a family, and even to eliminate a genetic disease from the human population.

So far, germline genetic modification of humans has remained a theory. Nevertheless, the idea has incited controversy. One ethical concern about germline genetic modification centers on consent. In the United States and many other countries, researchers must inform a potential subject about the risks posed by an experimental treatment. The researchers then must obtain the subject's voluntary consent to participate in the study. Studies involving germline genetic modification presents unknowable risks. A transgene may insert into chromosomes, resulting in disease or a lethal outcome. In addition, the technique affects future generations, who clearly cannot give their consent.

Religious commentators have expressed approval of somatic cell gene therapy, since its goal is to treat disease in individuals. However, this approval has not been given for germline genetic modification, which would, theoretically, treat a disease by altering the genomes of people as yet unborn. Critics say that advocates of germline genetic modification run the risk of "playing God." A related concern about the technique is that it could be used to manufacture humans to desired specifications, which might be a threat to principles of human dignity.

Opposition to germline genetic modification of humans run so strongly that Australia, Canada, Germany, and other countries have declared the technique illegal. Although the United States has not enacted a law to ban application of the method, the federal government will not financially support any research that aims to alter inherited genes of humans.

INDUCED PLURIPOTENT STEM CELLS

In 2006, physician and researcher Shinya Yamanaka at the Kyoto University in Japan created the first **induced pluripotent stem cell** from a mouse. With the induced pluripotent stem cell, differentiated cells can be reprogrammed so that they express genes that are normally active in embryonic stem cells and revert to a type of cell similar to early embryonic cells. In early studies, two dozen genes were used to produce induced pluripotent stem cells. Later, Yamanaka discovered that only four genes were required to change a differentiated cell into an induced pluripotent stem cell. Researchers have used induced pluripotent stem cells to correct diseases in experimental animal models. In 2007, human-induced pluripotent stem cells were produced by transforming human skin cells with four genes that integrated into the cells' nuclear DNA. Potentially, any human differentiated cell can be treated to become an induced pluripotent stem cell.

Inducing the production of pluripotent stem cells from a patient's own cells could provide another therapeutic approach that avoids immune system conflicts. However, studies with induced pluripotent stem cells are still in the early stages. Scientists warn that current techniques produce stem cells that may not be capable of fully differentiating into a desired cell type. The idea of using induced pluripotent stem cells for therapy has also stirred controversy because transgenic viruses are used to introduce genes into differentiated cells. Transgenic viruses have stimulated cancer growth in experimental animals. Therapies with induced pluripotent stem cells may be a decade away. Until then, induced pluripotent stem cells are used to produce cells representative of certain diseases. Scientists can study the cells to learn more about the diseases and can test new drugs before administering the drugs to patients.

HUMAN CLONING

In 1978, readers of the *New York Post* were greeted with the headline "Baby Born Without a Mother: He's First Human Clone." The article reported information from a forthcoming book, *In His Image: The Cloning of*

Man by science reporter David Rorvik. In his book, Rorvik told a story of how a California millionaire hired him to establish a lab on a Pacific Island where a science team labored for years before a surrogate mother gave birth to a healthy boy, an infant cloned from one of the millionaire's cells. Rorvik supported his amazing story with a description of research by Dr. J. Derek Bromhall, who had been a biologist at Oxford University. Bromhall had invented a method for performing nuclear transfer—the swapping of one nucleus for another nucleus—with mammalian cells. However, three years after publication, a judge decided that the book was a hoax; the book's publisher publicly apologized to Bromhall for including his genuine research in the fictitious story.

Despite its negative reputation, Rorvik's book sparked debates about the wisdom of cloning humans. Still, the controversy simmered until the cloning of Dolly the sheep was announced in 1997 and the culturing of human embryonic stem cells the following year. At that point, human cloning seemed possible, and stem cell research became mistakenly linked with fears about human cloning. A misunderstanding about the difference between **therapeutic cloning** and **reproductive cloning** forged this link.

Therapeutic cloning is a method used to produce stem cells for a patient. The procedure is performed by removing the nucleus from an egg cell, and replacing it with a nucleus removed from a somatic cell of a patient who needs stem cells. After nuclear transfer, the egg contains a nucleus with the patient's genome. The egg is allowed to divide and the cells develop to the early-stage embryo, the blastocyst. A scientist isolates cells from the blastocyst's inner cell mass to develop embryonic stem cell lines. Since the cells are pluripotent, they can differentiate into any cell of the body. As a method of producing embryonic stem cells for medical treatment, therapeutic cloning has the advantage that the stem cells should not provoke an immune reaction in the patient. These cells are tailored for a particular patient and should carry the patient's proteins on the cell surface. Those who oppose therapeutic cloning protest that the method requires the destruction of a human embryo.

Reproductive cloning would follow the same nuclear transfer procedure. Following nuclear transfer, however, the embryo would be implanted in a surrogate mother to create an infant that is a clone of the person who donated the nucleus. Most people vigorously oppose the theoretical technique of reproductive cloning. In July 2002, the President's Council on

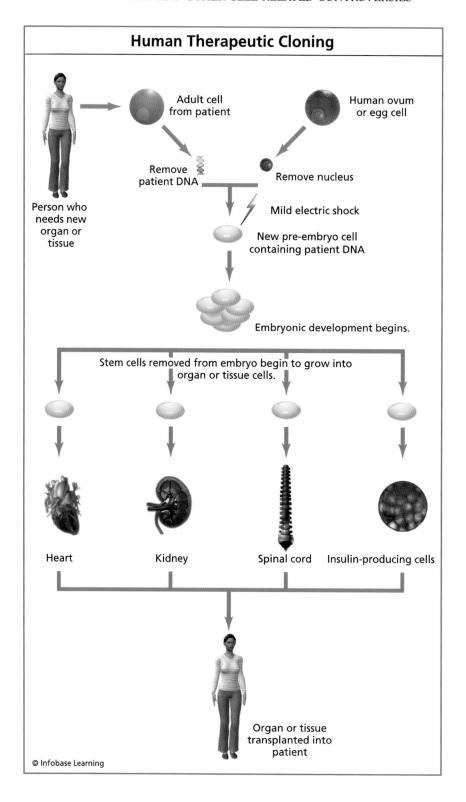

Genetic Modification of Humans 55

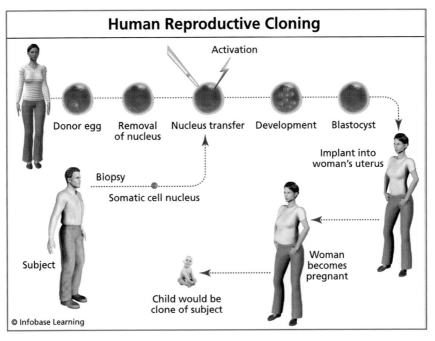

FIGURE 4.3 Human reproductive cloning is illegal in many countries.

Bioethics identified the following ethical concerns about reproductive cloning:

- Cloning would violate the principles of human research ethics because the technology is extremely unsafe, posing serious risks of birth defects and long-term health problems. Scientists may not be able to eliminate safety risks, the Council noted, because performing experiments with humans to make human cloning less dangerous would violate research ethics.
- Cloned children may experience problems of identity because each child would be virtually identical—genetically, at least—to a person who has lived his or her own life.
- If human cloning became an accepted practice, then society may view cloned children as manufactured products, which would threaten human dignity.
- In combination with germline genetic modification, human cloning could be used to pursue eugenics, an effort intended to improve

(Opposite page) FIGURE 4.2 Human therapeutic cloning could make it possible to create spare body parts that an individual's body would not reject.

DESIGNER GENES FOR ATHLETES?

For thousands of years, athletes have sought ways to enhance their performances beyond the physical training regimen. For example, ancient Greek athletes ingested stimulants to brace themselves for competition. Athletes of the nineteenth century dosed themselves with caffeine and cocaine. The practice became known as "doping," possibly from the Dutch word *dop*, which was the name of an alcoholic beverage consumed by Zulu warriors to enhance their strength and endurance in battle.

During the late 1990s, scientific groups showed that muscle performance of laboratory rodents can be improved with genetic engineering. This raised the concern that humans may undergo gene therapy to enhance their athletic performance. After all, some athletes had doped themselves with growth hormone proteins, which were produced by using recombinant DNA technology. Is gene doping so farfetched? The insulin-like growth factor 1 gene is one candidate for use in gene doping. One form of the growth factor stimulates an increase in muscle mass of transgenic mice. Increased muscle mass can also be achieved by blocking myostatin gene expression. The gene for erythropoietin offers yet another possibility for improving performance. Erythropoietin promotes the formation of red blood cells, and increased numbers of red blood cells enhance the transfer of oxygen to muscle tissue.

In 2001, the Medical Committee of the International Olympic Committee met to discuss the possibility of athletes' abuse of gene therapy. Two years later, the World Anti-Doping Agency banned the enhancement of athletic performance by a non-therapeutic use of genes or chemicals that alter gene expression. So far, cases of gene doping have not been detected. Experts suggest that abuse of gene therapies will be inevitable considering the sports community's interest in these techniques.

the human race. Aside from ethical problems, a eugenics program that promotes *genetic uniformity* would disregard the importance of *genetic diversity* for the survival of a species.

The Council concluded that reproductive cloning is unsafe and morally unacceptable. Opposition to this technology continues. When

President Obama lifted President Bush's restrictions on stem cell research in 2009, he promised that "we will ensure that our government never opens the door to the use of cloning for human reproduction. It is dangerous, profoundly wrong, and has no place in our society, or any society."

LIMITATIONS ON GENETIC ENGINEERING OF HUMANS

Safety concerns fuel controversy about somatic gene therapy, which is the transfer of a transgene into somatic cells. However, the risks appear to be diminishing as scientists and physicians gain more experience with gene therapy. Germline genetic modification of humans is another matter. The technology is currently unfeasible and considered to be ethically impermissible. It is even illegal in some countries. Similarly, the hypothetical cloning of humans has earned widespread condemnation. This does not mean that genetic modification of the human germline and cloning will never be practiced, but acceptance of the technologies will require a shift in society's values.

Property Rights in Genes and Tissues

THE PATENTING OF GENES AND CELLS

On May 9, 1873, the U.S. government granted French chemist and microbiologist Louis Pasteur a patent for a process of fermenting beer. This patent included a claim to yeast that was "free from organic germs of disease, as an article of manufacture." Pasteur's patent was one of the earliest U.S patents for a biological invention. More than a century later, the U.S. Supreme Court considered whether "living matter" is truly eligible for a patent. The court weighed in favor of patent protection. The landmark case, *Diamond v. Chakrabarty*, 447 U.S. 303 (1980), opened the doors for patenting of transgenic bacteria, transgenic animals, stem cells, DNA molecules, and other biotechnology inventions.

What is a Patent?

Property law gives the owner of personal property, such as a car, a computer, or a cell phone, the exclusive right to possess, use, and dispose of the property in any legal manner, and to prevent others from possessing the property. A patent has many characteristics of personal property. For example, a U.S. patent provides the patent owner, for a limited time, the right to prevent others from making or using the patented invention. A patent does not give the patent owner total control of the patented invention, because the right to make, use, or sell a patented invention may be regulated by federal, state, or local law. For example, the U.S. Food and

Drug Administration determines whether a patented drug can be marketed in the United States.

The rationale behind patent law is to encourage invention by guaranteeing certain property rights for inventions. In return for limited property rights, the inventor must file a patent application that describes the invention with the U.S. Patent and Trademark Office (USPTO). A patent can reward inventors and their financial backers to justify the time and money they risk in research and development efforts. At the same time, the patent system encourages public disclosure of technical information, which may otherwise have remained secret, so that others can use the information to improve the invention or to devise different inventions based on data in a patent.

A **utility patent** is the most common type of patent sought by those who work in the field of biotechnology. The U.S. government grants a utility patent for new and useful inventions. To obtain a utility patent, an inventor must file a patent application with the USPTO that fully explains the invention, including detailed methods for making and using it. The patent application also must include claims that precisely describe the invention. In real estate, a deed contains a section that sets out the boundaries of the land. Similarly, patent claims set out the limits of the invention. As an example, the first claim of U.S. Patent No. 6,498,285 describes an invented method for producing a transgenic pig:

> A method for producing a transgenic pig comprising: a) obtaining a porcine embryo comprising at least three blastomeres; b) introducing at least one clone of isolated nucleic acid molecules into at least one blastomere of the embryo; c) transferring the embryo to a surrogate female pig; d) developing the embryo into at least the fetal stage; and e) developing the fetus into a transgenic pig.

After an inventor files a patent application with the USPTO, a patent examiner scrutinizes the application to determine if it meets three types of requirements for a patent:

- The patent claims must define something that can be patented under U.S. law. A U.S. patent may claim a process, machine, manufacture, or composition of matter. Basically, anything that results from creative, human action may be patentable. Laws of nature, physical phenomena, and abstract ideas are not patentable.

- The patent claims must not describe something that is already known; that is, the claimed invention must be new. The claimed invention also must not be an obvious variation of something that is known.
- The patent application must adequately describe the invention.

If a patent examiner decides that a patent application fulfills these requirements, then the examiner should grant a patent. During the life of a patent, a patent owner has the right to prevent someone from making, using, selling, or even offering to sell the patented invention within the United States. A patent owner can also stop a person from importing the patented invention into the United States. These rights can be enforced by suing for patent infringement. Patent rights last for 20 years from the filing date of the patent application. Since the patent examination process can require two, three, four or more years, a patent term tends to be much less than 20 years.

Biotechnology Patents: Examples and Controversies

The U.S. Supreme Court case, *Diamond v. Chakrabarty*, set the stage for the era of biotechnology patents. The *Chakrabarty* case began in 1972 when Ananda M. Chakrabarty, then a research microbiologist at General Electric Company, invented a type of bacteria that would degrade two or more of the main components of crude oil. Chakrabarty developed the bacteria by isolating plasmids from different strains of bacteria, each of which gave an original bacteria strain the ability to degrade one oil component. He then inserted at least two plasmids into a single bacterium. The new strain of bacteria was designed to be placed on an oil spill to break down the oil into harmless products. After the bacteria depleted the oil, the microorganisms died. Chakrabarty filed a patent application containing claims to the method of producing the bacteria, and claims to the new type of bacteria.

A USPTO examiner rejected claims to the bacteria, asserting that the bacteria were "products of nature" and, therefore, cannot be patented under U.S. patent law. The case eventually landed in the U.S. Supreme Court, which ruled that Chakrabarty's bacteria could be patented. The Court decided that the inventor had produced a new type of bacteria with markedly different characteristics from any found in nature. "His discovery is not nature's handiwork," the Court said, "but his own."

The same principle applies to molecules, such as proteins, DNA, and RNA. A naturally occurring molecule cannot be patented, even if the patent applicant was the first to discover the existence of the molecule.

FIGURE 5.1 Ananda Mohan Chakrabarty (*pictured here*) genetically engineered a new species of oil-eating bacteria in 1971, while working for General Electrical Company in Schenectady, New York, and then applied for a patent on a living organism.

Alternatively, a purified, isolated, or altered form of a naturally occurring molecule may be patentable. For example, an early court case on gene patenting concerned claims to DNA molecules that encoded human erythropoietin, a protein that increases the formation of red blood cells. A federal judge explained that the nucleotide sequence encoding human erythropoietin is not patentable because the erythropoietin gene is a natural phenomenon. However, an *isolated and purified* DNA molecule that encodes human erythropoietin can be patented. The isolated and purified DNA molecule does not exist in nature.

The idea of patenting genes and DNA molecules has incited controversies at least since the early 1990s when the NIH filed three patent applications that covered more than 6,000 DNA fragments, which may be used to find genes. James Watson, co-discoverer of DNA's double helix structure, denounced the plan to patent DNA bits as "sheer lunacy." Watson declared that "virtually any monkey" could crank out the nucleotide sequences using automated sequencing machines. Facing worldwide criticism, the National Institutes of Health withdrew the patent applications.

Robert Cook-Deegan, Director of Duke University's Center for Genome Ethics, Law and Policy, has identified three types of gene patenting disputes that raise legal and ethical concerns:

- *Therapeutic proteins*: Many patent claims describe methods and DNA molecules used to produce proteins for treatment of human diseases. These patents inspire lawsuits among pharmaceutical and biotechnology firms over patent rights. Companies fight these legal battles to secure patent rights in top-selling therapeutics.
- *Scientific Research*: Many protest against gene patents on the grounds that the patents could hinder scientific progress. University researchers, for example, might avoid investigating a gene because of a company's patent claims to that gene.
- *Diagnostics*: As discussed earlier, genetic mutations can create an increased risk of developing a disease. The claims of certain patents cover DNA molecules that can be used to identify a mutation, and some patent claims cover DNA screening tests for genetic mutations. Those who oppose genetic test patents contend that these patents raise gene test prices and discourage others from improving patented gene tests.

BATTLE FOR THE HALF-HUMAN HYBRIDS PATENT

Professor Stuart Newman of New York Medical College filed a patent application in 1997 for a bizarre invention: half-human hybrids. Newman described a method for combining human embryo cells with embryo cells obtained from a monkey, a pig, a mouse, or other non-human animal. The hybrids would be useful for biological research and as a source of tissue for transplants, the inventor proclaimed.

A USPTO examiner, who asserted that the human hybrid claims included humans, rejected the claims on the basis that they failed to meet many of the legal requirements for patentability. The examiner highlighted a USPTO policy published in the late 1980s when the idea of patenting non-human transgenic animals created a controversy. The USPTO had announced that a patent claim that includes a human being would not qualify for patent protection because such a claim would limit the rights of a human, and this would be prohibited by the U.S. Constitution.

Nevertheless, Newman pursued his patent claim for seven years. In 2004, the examiner informed Newman about a recently enacted federal law banning the granting of patents with claims that included humans. The inventor conceded his defeat, but then turned around to celebrate it as a victory. After he ended his struggle for the human hybrid patent, Newman admitted that he had never wanted the patent in the first place. He opposed patents on living things and wanted to create a legal precedent to prevent the USPTO from granting patents on human hybrids.

Nevertheless, Newman's battle with the patent office leaves an important issue unresolved: How many human genes and human cells would a transgenic animal need to contain before it could be considered as deserving of human rights? As USPTO deputy commissioner John Doll told *The Washington Post*, "I don't think anyone knows in terms of crude percentages how to differentiate between humans and nonhumans." The USPTO could use guidance from the courts or Congress, Doll said.

Cook-Deegan concluded that gene patents have supported the development of some therapeutic proteins and that, so far, gene patents have not hindered scientific research. He also said that any harms or benefits of gene-testing patents remain unclear.

Nevertheless, ill will against gene-testing patents persists. The most famous, controversial case concerns patents on DNA molecules used for encoding BRCA1, a protein linked to susceptibility for breast cancer. Disputes about the patents have endured for more than a decade. In a recent tactic developed to eliminate the patents, the Association for Molecular Pathology and others sued the USPTO and patent owners in 2009. Their suit alleged that patent claims to molecules with BRCA1 nucleotide sequences are invalid. U.S. District Judge Robert W. Sweet agreed. In his 2010 decision, the judge showed sympathy for those who protested that the claiming of DNA as isolated DNA molecules is a "lawyer's trick," which avoids the ban "on the direct patenting of DNA in our bodies but which, in practice, reaches the same result." The case is on appeal and may reach the U.S. Supreme Court.

CONTROVERSIES ABOUT OWNERSHIP OF BIOLOGICAL TISSUE

The guild known as the Company of Barber-Surgeons held the exclusive right to perform human anatomy demonstrations in Great Britain for 175 years. By the eighteenth century, private schools of human anatomy appeared. These new schools promoted a different approach than that used by the Barber-Surgeons guild: Students were permitted to dissect a human body to gain hands-on experience. With the new anatomy schools, the need for corpses skyrocketed. Professional body snatchers fulfilled this demand. Known as "Sack-Em-Up Men," "Resurrectionists," or "Resurrection Men," they robbed graves for profit.

Working at night, the Resurrection Men dug up corpses, carefully stripped them, and then tossed the clothes and personal effects back into the grave. The theft of clothing or other items from a grave was a crime. The body, however, was not considered to be property and, therefore, it could not be owned or stolen. Surprisingly, the years brought little change to the laws concerning the ownership of a body.

Who Owns Biological Tissue Samples?

According to a 1999 Rand Corporation study, facilities within the United States stored at least 307 million tissue specimens, an amount that increases at a rate of more than 20 million samples a year. Most of the stored tissues were collected to diagnose or to treat a disease. After tissue samples are no longer needed for these purposes, they are often used for research or education. Who owns these tissues? Do individual donors have any rights to excised pieces of their own bodies?

One of the most famous cases concerning tissue ownership began in 1976 when John Moore visited the University of California, Los Angeles Medical Center, following a diagnosis of leukemia, a cancer of the blood cells. Doctors removed blood, bone marrow, and other tissue samples to confirm the diagnosis. While studying Moore's tissues, doctors realized that the cells might be useful for academic and industrial research. A doctor recommended the surgical removal of John Moore's spleen to slow the progression of the cancer. Before the doctors discarded the excised spleen, they removed a tissue sample for research, and, over the next three years, established a cell line of white blood cells. The doctors obtained a patent on the cell line in 1984, and the University of California acquired revenue from companies that wanted to use the patented cells.

When Moore learned about the cell line and its commercialization, he sued on 13 grounds, including a claim that he had an ownership interest in the patent. The case eventually landed in the Supreme Court of California. The judges concluded that Moore lacked an ownership interest in the patent because he was not one of the inventors of the cell line. Moore also alleged that the doctors were liable for conversion, which means that they had interfered with Moore's rights to posses and own personal property. However, the judges decided that Moore did not have any ownership interest in his cells after their removal from his body. In addition, applying conversion law to cover rights in tissue samples would hinder medical research. The court did rule in Moore's favor on one point: Moore could sue his doctor for breaching a physician's duty to inform a patient of any economic or personal interest in using his tissues. A patient should be aware of such interests, which may affect a doctor's medical judgment.

Since the *Moore* decision, other courts have considered property rights in excised tissues. For example, a Florida court in 2003 reviewed a case concerning Canavan disease, which is an inherited disorder of the nervous system. Canavan disease is caused by a mutation in a gene that encodes the enzyme aspartoacylase. The mutation results in a lack of the enzyme's activity, which, in turn, causes a buildup of a chemical (N-acetylaspartic acid) that damages nerve cells. The case involved doctors and scientists, who used donated tissue samples to isolate the gene that causes Canavan disease. After they obtained a patent on the gene, the tissue donors sued for conversion, among other grounds. The court decided that a person has no property interest in tissue or genetic material donated for research. In 2007, yet another court, a federal appellate court, concluded that patients had no property interest in their tissues after they had donated the material as a gift for medical studies.

Hospitals and universities continue to struggle with the question of the ownership of human biological samples. Organizations can request patients to sign informed consent documents that state the possibility that their tissues may be used for basic research. However, ensuring a patient's informed consent about future uses of tissue removed for diagnosis or treatment raises problems of its own. Individual patients may consent to certain types of research with their tissues, but what happens when new discoveries lead researchers to areas not contemplated in an informed consent form? Is it necessary to track down a patient and get a signature on a new consent document? As medical organizations ponder the question of informed consent, courts avoid the question about whether people own their tissues. Instead, judges try to strike a balance between the promotion of research with human tissue and respect for an individual's sense of freedom and independence.

HUMAN EMBRYONIC STEM CELL RESEARCH: PATENT DISPUTES AND TISSUE OWNERSHIP ISSUES

Human embryonic stem cell research provokes controversies about patent rights. Embryonic stem cell patents exist in a "complex and troubled landscape," wrote Aurora Plomer, Kenneth S. Taymor, and Christopher Thomas Scott, experts in biomedical ethics, law, and business, in a 2008 issue of *Cell Stem Cell*. "The patenting and commercialization of hES [human embryonic stem] cells have produced one of the most unusual and fraught situations in the history of science, ethics, and law," they said. "hES cell intellectual property [i.e., patents] and its future impact on human health will be guided and bound by a welter of moral, technical, and legal inputs...."

A series of legal battles, for example, focus on three key human embryonic stem cell patents, which name the University of Wisconsin's James A. Thomson as inventor. Consumer groups targeted the Thomas patents for destruction in skirmishes with the patent office. Following a two-year fight, the USPTO reapproved the patents. However, the protestors did not give up. In 2010, they convinced the USPTO Board of Patent Appeals to send one patent back to an examiner to reconsider the validity of the patent claims. "This is great news for medical research," said

BIOPIRATES?

In his book, *The Thief at the End of the World* (2008), author Joe Jackson tells the story of adventurer Henry Wickham, who returned to England from Brazil in 1876 with a secret cargo: 70,000 rubber tree seeds. The stolen seeds were planted in Britain's Far East colonies. By the early twentieth century, the British Empire dominated world rubber production. Wickham's theft of seeds was consistent with a traditional view that biological resources are the common heritage of mankind and can be removed from any country without breaking a law, without the need to obtain authorization, and without any need to compensate for taking the resource. As Jackson wrote, the modern view is that Wickham performed an act of biopiracy.

Biopiracy is the unauthorized and uncompensated taking and use of a biological resource for a commercial purpose. Claims of biopiracy are often heard when a business from a developed nation claims ownership and patents traditional knowledge of a developing nation, or patents organisms and genes discovered in a developing nation without the consent of, or compensation to, local governments.

One biopiracy controversy began when the European Patent Office granted a patent that covered a method of using neem tree oil to kill fungus. The neem tree is common in India, where the plant has been used for centuries as a source of medicine and other useful compositions. Various groups protested the patent, and the European Patent Office reconsidered. At a hearing, the manager of an Indian agriculture company showed that he used neem tree oil as an anti-fungal agent years before the patent was filed. Since the patent claimed a method that was not new, the European Patent Office revoked the patent.

During 2010, participants of the United Nations Convention on Biological Diversity worked on a global agreement about biopiracy. Their objective is to create a way to regulate access to genetic resources and the distribution of profits resulting from commercial use of resources.

Dr. Jeanne Loring, director of the Center for Regenerative Medicine at the Scripps Institute in San Diego, California. "Human embryonic stem cells hold great promise for advancing human health, and no one has the ethical right to own them."

Stem cell research also raises issues about property rights in donated tissue. When President Obama removed limitations on human stem cell research in 2009, he also directed the National Institutes of Health to issue guidelines for federal funding of research with human embryonic stem cells. The new guidelines state that human embryonic stem cells should have been derived from human embryos that were created by in vitro fertilization for reproductive purposes and were no longer needed for this purpose. The donors of the embryos must provide written consent for the embryos to be used for research. While U.S. laws do not establish any ownership rights in donated tissues, donors who sign the informed consent form do have some rights: Donors can withdraw consent for embryo donation until researchers use the embryos to derive stem cells.

6
Medical Genetic Testing

A **genetic test** can either examine the function of a gene or determine the presence of a gene mutation. Three types of genetic tests can reveal whether a person has a genetic disease:

- Tissue sample analysis for a particular RNA or protein may show that a person's cells synthesize a protein associated with a disorder, or do not synthesize a protein required for good health.
- A microscopic study of cells can reveal the presence of altered chromosomes.
- DNA can be isolated from a tissue sample and analyzed for the presence of a mutated gene.

Laboratories perform more than 1,000 different genetic tests. Traditionally, physicians have used laboratory test results to diagnose a disease and to determine a prognosis for a patient. In addition, genetic tests can indicate the risk of developing a disease, the chances that a child will inherit a disease gene, and whether a particular medicine may cause more harm than good.

DETECTING DIFFERENCES IN DNA NUCLEOTIDE SEQUENCES

During the late 1970s, Dr. Alec Jeffreys, a geneticist at the University of Leicester in the United Kingdom, discovered the presence of certain

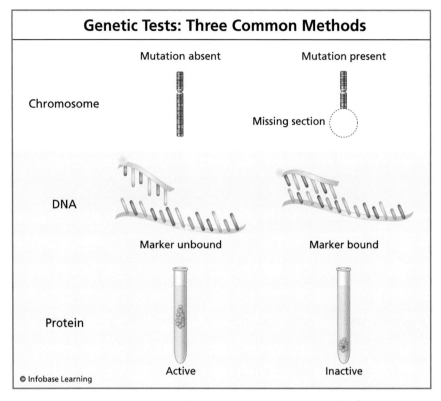

FIGURE 6.1 Genetic testing involves examining tissue, typically from cells in a blood sample, for mutations linked to a disease or disorder. Some tests identify changes in whole chromosomes. Others examine short sections of genes, or look for the protein products of genes.

alterations in the nucleotide sequences of human DNA, which affect the activity of restriction enzymes. Recall that a restriction enzyme is an enzyme that cuts DNA at a specific place. A restriction enzyme moves along a DNA molecule until it comes across its cleavage site. Then, the enzyme binds to the DNA molecule and breaks the DNA backbone. An alteration in a nucleotide sequence can either abolish a cleavage site or create a new site. As a result, restriction enzyme digestion of DNA from two individuals can produce DNA fragments of different sizes. The term *polymorphism* refers to something, such as DNA, that exists in different forms, and a "length polymorphism" is something that exists in different lengths. Since digestion of DNA by restriction enzymes produces DNA fragments and reveals length polymorphism of the DNA fragments, the technique is called **restriction fragment length polymorphism (RFLP)**. Simply put,

RFLP relies upon the ability of restriction enzymes to cleave non-identical DNA samples into fragments that have different lengths.

RFLP is performed by digesting DNA using restriction enzymes to produce DNA fragments, and then separating the DNA fragments by size using electrophoresis. Electrophoresis is carried out by placing enzyme-treated DNA samples in a gel slab. An electric current pulls the negatively charged DNA fragments through the gel and toward the positive electrode at the other end of the gel. The gel acts like a sieve: Smaller DNA fragments tumble through the gel maze faster than larger DNA fragments. Electrophoresis separates DNA fragments of different size and creates a pattern of DNA fragments that looks like a bar code. Following electrophoresis, the presence of a mutated gene in a DNA sample can be detected using a radioactively labeled molecule that binds with the nucleotide sequence of the gene.

Over the years, researchers have refined RFLP analysis by taking advantage of the **polymerase chain reaction (PCR)** to synthesize many copies of the gene of interest. A DNA polymerase is an enzyme that

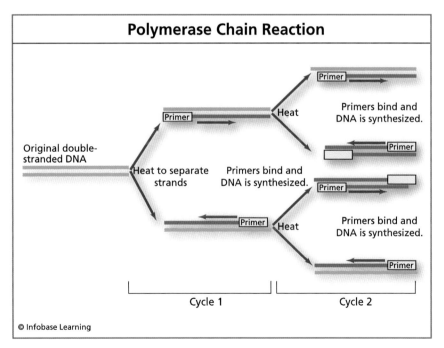

FIGURE 6.2 The polymerase chain reaction quickly produces a large number of copies of a nucleotide sequence. The process shown here continues into further cycles.

synthesizes DNA from nucleotides. Scientists perform the PCR technique, which is sometimes called molecular photocopying, with a DNA polymerase that can resist high temperatures. Before carrying out the PCR technique, a scientist must decide which nucleotide sequence to copy. The selected nucleotide sequence is sometimes called the target nucleotide sequence. The scientist makes two types of DNA molecules called primers. A primer is a single-stranded DNA molecule that has about 20 nucleotides. The primers have nucleotide sequences that allow them to bind with segments of DNA found at both ends of the target nucleotide sequence.

To perform the PCR technique, a researcher heats a sample of DNA to separate the strands of double-stranded DNA molecules. The researcher cools the sample and adds the two types of primers. Each primer binds to a DNA molecule strand at one end of the target nucleotide sequence. Then, DNA polymerase synthesizes DNA by adding nucleotides to the primers. While copying DNA, DNA polymerase adds the correct nucleotides according to the nucleotide rules of attraction: T pairs with an A, and

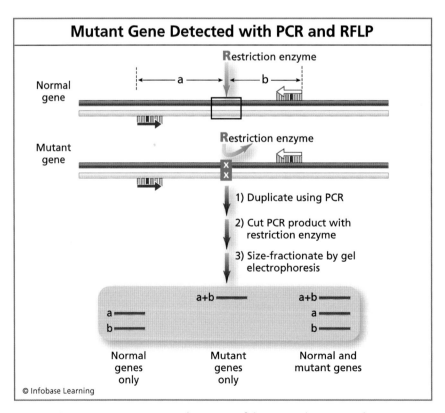

FIGURE 6.3 Restriction enzyme digestion of the normal gene produces two fragments, whereas digestion of the mutant gene produces one fragment.

Medical Genetic Testing

G pairs with a C. After completing DNA synthesis, the DNA sample is heated again to separate DNA strands. Cooling the sample allows primers to bind with single-stranded DNA molecules. Another round of synthesis begins. By repeating these steps, a technician can make a billion copies of the target nucleotide sequence in a few hours.

An Example of a DNA Test for a Gene Mutation

Researchers have identified a mutation in the Wolfram syndrome (WFS1) gene, which is associated with a type of hearing loss. This mutation abolishes a cleavage site for the restriction enzyme *Hae*II, which recognizes the cleavage site, "RGCGCY," where R can be A or G, and Y can be C or T. A segment of the normal WFS1 gene has the following nucleotide sequence with a *Hae*II cleavage site, which is shown in red in the top DNA strand:

.... CAAC**AGCGCC**GAGTCT ....

.... GTTGT**CGCGG**CTCAGA ....

Notice that the bottom DNA strand also has a *Hae*II cleavage site (shown in blue). Since it lies in the opposite DNA strand, the cleavage site reads in the opposite direction. The enzyme *Hae*II binds to its cleavage site in this area of the WFS1 gene and cuts the DNA molecule as shown below.

.... CAAC**AGCGC**   CGAGTCT ....

.... GTTGT   CGCGGCTCAGA ....

A mutation results in the substitution of A for G in this segment of the WFS1 gene.

Normal gene: CAACAGCGCCGAGTCT

Mutant gene: CAACAGCACCGAGTCT

This mutation abolishes a cleavage site for *Hae*II.

To test for the presence of the mutated WFS1 gene, researchers used the PCR technique to synthesize 573 base pair DNA fragments that include the nucleotide sequence with a possible mutation. A 573-base pair DNA fragment from a normal WFS1 gene has two *Hae*II cleavage sites at positions 242 and 455.

                                       242   455

                                       ↓   ↓

573 base pair DNA fragment: ─────────────────

JUMPING GENES IN THE HUMAN GENOME

In October 1990, scientists launched the Human Genome Project, a colossal, international effort to learn the sequence of the human genome's three billion base pairs. Researchers sequenced parts of genomes of different people, collected the results, and assembled the data like pieces of a puzzle. Early results indicated that humans overall share an identical order of 99.9% of their nucleotide bases.

New technologies enable researchers to investigate the genomes of individuals. Scott E. Devine of the University of Maryland School of Medicine led a research team who examined the genomes of 76 people. They discovered that human genomes have more variation than expected from percent identity numbers due to surprisingly common transposons. Also known as "jumping genes," transposons are DNA segments that duplicate themselves and move to new areas in a person's genome. Since transposons accumulate, the genomes of each generation of a family include more transposons that the genomes of the family's ancestors.

As they insert themselves in various sites of a genome, transposons can disrupt normal gene function. "If you think of the human genome as a manual to build a complex machine like an aircraft," Devine explained in a press release, "imagine what would happen if you copied the page that describes passenger seats and inserted it into the section that describes jet engines. Transposons act something like this: They copy themselves and insert the copies into other areas of the human genome, areas that contain instructions for the complex machine that is the human body." Transposon insertion can corrupt the nucleotide sequence data of normal genes. "This, in turn," Devine said, "can alter human traits or even cause human diseases." Devine's team found that some transposons appear to be associated with lung cancer.

In contrast, a DNA fragment from a mutant WFS1 gene has only one *Hae*II cleavage site at position 455. This means that *Hae*II treatment of a 573-base pair DNA molecule from a normal WFS1 gene yields DNA fragments of 242, 213, and 118 base pairs, whereas *Hae*II treatment of a 573-base pair DNA molecule from a mutant WFS1 gene yields DNA fragments of 455 and 118 base pairs. The scientists who invented the test observed that DNA samples from some people resulted in four sizes of DNA fragments: 455, 242, 213, and 118 base pairs. This result shows that an individual has one copy of the normal gene, as well as one copy of the mutant gene, inherited from the mother or father.

TESTING FOR GENES AND GENE FUNCTION RELATED TO A DISEASE

When a patient describes certain symptoms, a doctor may order a genetic test to confirm a diagnosis of a disease at an early stage. For example, polycythemia vera is a rare blood disorder that may produce symptoms in the early stages that suggest many common health problems. If polycythemia vera is left untreated, it can cause death within two years. A mutation in the JAK2 gene occurs in up to 97% of patients with a confirmed diagnosis of polycythemia vera. A genetic test for this mutation performed during the early stage of a patient's disease can help a doctor to treat the patient in time. In addition to confirming diagnosis, genetic tests are performed for many reasons, including

- *carrier testing*: Individuals may request a genetic test to determine whether they are a carrier for a genetic disease. A carrier is a person who has inherited one copy of a disease gene from either the mother or father. Carrier testing may reveal that a woman and man are carriers for a disease gene. If so, then their child may inherit a copy of the disease gene from both parents. Some disorders require two copies of a disease gene before medical problems become severe. Cystic fibrosis, for example, occurs when a person inherits a mutated gene from both parents.
- *pre-implantation diagnosis*: In the process of in vitro fertilization, egg and sperm cells combine to form an embryo, which will be implanted into a mother's uterus. A couple may decide to request genetic tests for disease genes in newly formed embryos. The test may require the removal of one cell from each eight-cell embryo. Pre-implantation genetic diagnosis can provide an opportunity for a fertility specialist to select embryos that lack disease genes before implantation.
- *prenatal testing*: Carrier testing or family histories may indicate that a couple has a high risk of conceiving a child with a certain genetic disease. If so, then the couple may request prenatal testing, which tests the fetus for the presence of the disease. Individuals

of advanced age may also decide to request prenatal testing. In North America, pregnant women who are older than 35 years often request an analysis of fetal chromosomes to detect Down's syndrome and other disorders. One method of prenatal testing is amniocentesis, in which a doctor inserts a hollow needle into the woman's uterus and takes a small amount of the fluid—the amniotic fluid—that surrounds the fetus. The fluid contains fetal cells, which can be tested.

- *newborn screening:* All states in the United States require the genetic testing of healthy-appearing newborns to detect disorders that have yet to produce symptoms. Newborn screening is one of the most extensive U.S. public health activities. A test to detect phenylketonuria is one example of a newborn test. An infant who is found to have this disorder can be put on a restricted diet to prevent the development of mental retardation.
- *predictive genetic tests:* These tests can reveal whether a currently healthy person has a risk of developing a certain disease in the future. A positive test for some diseases can lead to treatment that prevents the disease. However, other diseases cannot be cured. For instance, Huntington's disease is an inherited disorder that causes degeneration of brain cells. Although treatments can alleviate some symptoms, scientists have not yet devised a cure for Huntington's disease. A genetic test can detect the presence of this disease gene many years before symptoms appear.
- *genetic risk factor tests:* A search for genetic risk factors can identify a person's risk for developing a complex disease, such as cancer, diabetes, or heart disease. In these cases, genes do not determine that a particular individual will develop a certain disease, but rather, certain genes make that person more susceptible to developing a disease. Doctors can use data about genetic risk factors to help their patients maintain good health.
- *personalized medicine:* Certain genetic variations influence health by affecting how a person reacts to a drug. Many of these variations occur in genes that encode enzymes that break down drugs. Some enzymes, for example, alter a drug to make it more soluble in water. As a result, the body excretes the soluble drug, which decreases the amount of drug that remains in a person's bloodstream. Variations in the activity levels of drug-altering enzymes determine the effectiveness of a drug and whether a

person will experience toxic side effects of the drug. Scientists are devising tests that reveal variations in the genes of enzymes that alter drugs. These efforts create the basis for "personalized medicine" wherein a doctor can tailor a drug treatment for an individual that ensures its effectiveness and reduces harmful side effects. The administration of the drug warfarin to prevent blood clots is one example of this approach. Genetic variations in several enzymes complicate warfarin dosage. A patient may require many cycles of trial and error during the first year of treatment, which can place a patient at risk for excessive bleeding or blood clots. The FDA recommends genetic tests to avoid the dangerous effects of an inappropriate dose of warfarin.

CONCERNS ABOUT GENETIC TESTS

Genetic testing has ignited debates, especially arguments surrounding pre-implantation and prenatal testing. Some critics claim that couples

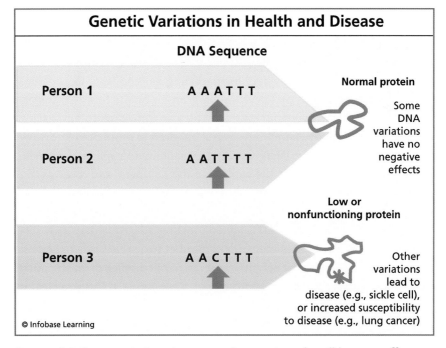

FIGURE 6.4 Some variations in a person's genetic code will have no effect on the protein that is produced, but others may lead to an increased susceptibility to specific diseases.

could use pre-implantation tests to select for cosmetic and physical traits. A prenatal genetic test may reveal that the fetus has an untreatable and devastating genetic disease, a result that may lead to a decision to end the pregnancy. Such a decision offends those who are opposed to abortion. Even more controversial is a decision to end pregnancy if a fetus has a treatable disease, a disorder that causes a mild cosmetic defect, or if the parents decide that the fetus is of the wrong gender.

Predictive genetic tests are controversial when they can reveal that a currently healthy person will probably develop a disease late in life for which no treatment currently exists. A laboratory may refuse to perform a predictive genetic test until the individual receives genetic counseling and provides informed consent to the procedure. Some people decide that it is

DTC GENETIC TESTS: BUYER BEWARE

Typically, a physician, genetic counselor, or other health-care provider orders a genetic test for a patient. The health-care provider arranges for the collection of a patient's tissue sample, which is sent to a laboratory for analysis. After reviewing the test results, the health-care provider interprets the findings for the patient. In recent years, companies have bypassed health-care providers by marketing genetic tests to consumers via advertisements on television, in print, or on the Internet. Direct-to-consumer (DTC) genetic testing enables an individual to purchase a test kit—costing anywhere from several hundred to more than $1,000—and collect a sample of DNA, usually by swabbing the inside of a cheek. The consumer then mails the sample to the company and receives the results by a telephone call or a letter. Sometimes, companies post the results online.

Companies that offer DTC genetic testing advertise that their services can diagnose genetic disorders, predict toxic reactions to certain drugs, and estimate susceptibility to a range of complex diseases, such as cancer and diabetes. Dr. Nancy Press, a professor at Oregon Health and Science University, says that consumers should be cautious about such promises. A negative result for a genetic disease gene may falsely reassure a consumer, who delays doctor visits or routine screening for a disorder. Press also says that genetic tests advertised to estimate better not to know, while others prefer to know that they will face a devastating disease so that they can make important decisions, such as career choice, marriage, and whether or not to have children.

One general dispute in the area of genetic testing focuses on the use of an individual's genetic data. An employer may want access to such information to hire employees who have a low risk of developing a disease. Health and life insurance companies would also benefit from accessing genetic information before agreeing to insure a person. In the United States, Congress has attempted to protect a person's genetic data with the Genetic Information Nondiscrimination Act, which forbids insurers from denying coverage or charging higher premiums based solely on a genetic tendency to develop a disease in the future. The act also prohibits

susceptibility to complex diseases are insufficient to predict the development of a disease.

Some companies offer "nutrigenomics" tests, in which genetic test results are supposedly used to customize a nutrition program for the customer. The FDA and the Centers for Disease Control and Prevention assert that their experts do not know about any valid scientific studies indicating that genetic tests can be used to effectively recommend nutritional choices. Perhaps this explains an observation by the Federal Trade Commission that what customers usually receive from these companies is advice that resembles standard sensible dietary recommendations.

The U.S. Government Accountability Office (GAO) issued a 2010 report about its investigation into DTC genetic testing companies. In one part of the study, GAO staff posed as consumers, bought genetic tests from four companies, and sent DNA samples donated by volunteers. "GAO's fictitious consumers received test results that are misleading and of little or no practical use," reported Gregory Kutz, managing director of special investigations at the GAO. "For example, GAO's donors often received disease risk predictions that varied across the four companies, indicating that identical DNA samples yield contradictory results." Kutz also said that GAO donors "received DNA-based disease predictions that conflicted with their actual medical conditions."

employers from basing an employment decision upon a job applicant's genetic information. States have enacted various genetic privacy statutes that require a person's consent to disclose their genetic test results, prohibit insurers from denying or restricting coverage based on an individual's genetic test results, forbid employers from requiring a job applicant to get a genetic test, and from using genetic data to deny employment.

A CAUTION ABOUT GENETIC RISK FACTOR TESTING

In the United States, the Human Genome Project was coordinated by the National Institutes of Health and the Department of Energy, which had expertise in genetic research pertaining to the effects of radiation and chemicals on human health. The Department of Energy offers educational information to the public about certain uses of genetic data. "Gene testing already has dramatically improved lives," states the Web site of the U.S. Department of Energy Genome Programs. "Some tests are used to clarify a diagnosis and direct a physician toward appropriate treatments, while others allow families to avoid having children with devastating diseases or identify people at high risk for conditions that may be preventable."

According to the federal agency, genetic tests for Alzheimer's disease and other complex, adult-onset disorders have provoked much of the current debate about testing genes. "The tests give only a probability for developing the disorder," the agency stresses. "One of the most serious limitations of these susceptibility tests is the difficulty in interpreting a positive result because some people who carry a disease-associated mutation never develop the disease. Scientists believe that these mutations may work together with other, unknown mutations or with environmental factors to cause disease."

Genetic risk factor tests may represent a case where a technology exists ahead of its time. "Many in the medical establishment feel that uncertainties surrounding test interpretation," the Department of Energy Genome Programs say, "the current lack of available medical options for these diseases, the tests' potential for provoking anxiety, and risks for discrimination and social stigmatization could outweigh the benefits of testing."

Forensic DNA Analysis

As discussed in the previous chapter, British geneticist Dr. Alec Jeffreys discovered that alterations in the nucleotide sequences of human DNA affect the activity of restriction enzymes. A change in a nucleotide sequence can create a cleavage site for a particular restriction enzyme or abolish a cleavage site. During the early 1980s, Jeffreys investigated certain regions of human DNA—known as minisatellites—that do not contain genes but do include nucleotide sequences repeated end to end, often up to hundreds of times. Picture a repeated nucleotide sequence as a boxcar of a train. A segment of DNA containing nucleotide sequence repeats would be a line of identical boxcars. The number of boxcars connecting the engine and the caboose determines the length of the train. Similarly, the number of repeated nucleotide sequences determines the total length of DNA containing the repeated sequences. Since the number of repetitions within any particular minisatellite varies among individuals, distances between the nucleotide sequences at either end of one of these nucleotide sequence stutters also vary. Jeffreys visualized these length polymorphisms using the restriction fragment length polymorphism (RFLP) technique.

Jefferys' investigation of minisatellites led him to develop a method for distinguishing DNA from two individuals. Soon after he published a report on DNA fingerprinting, Jeffreys applied the technique in an immigration dispute by showing that 15-year-old Andrew Sarbah, who had returned to the United Kingdom (UK) after 11 years in Ghana, was the

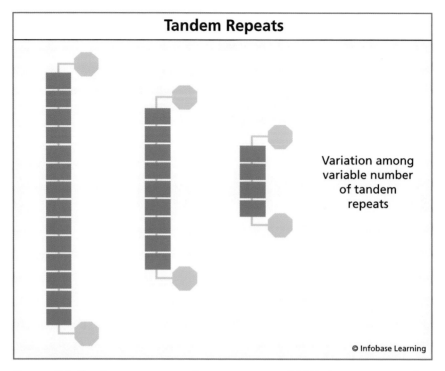

FIGURE 7.1 Tandem repeats are short sequences of DNA that are repeated top-to-bottom at a specific point. They are found throughout the human genome.

son of a UK citizen. Within a year, DNA fingerprinting sparked a revolution in criminal investigation.

In 1986, police discovered a rapist-killer's second victim in the village of Narborough in Leicestershire, England. When the investigation focused on a teenage boy as the main suspect, the Leicestershire constabulary asked Jeffreys to perform DNA fingerprinting to compare DNA extracted from the suspect's blood sample with the killer-rapist's DNA recovered from the crime scenes. Jeffreys showed that the boy was not the murderer. He also verified that one man was responsible for both killings.

The police decided to flush out the killer by collecting blood samples for DNA testing from every local male between the ages of 16 and 34. During this massive effort, a woman who worked at a Leicester bakery informed the police that a coworker had donated blood on behalf of a cake decorator named Colin Pitchfork. The police questioned and arrested Pitchfork, who confessed to the crimes of rape and murder. DNA fingerprinting confirmed that Colin Pitchfork's DNA (the 4,583rd DNA sample tested) and the killer-rapist's DNA shared identical profiles. Since then, DNA analysis emerged as an essential tool of law enforcement.

DNA PROFILING

In the United States, the RFLP technique became the first scientifically accepted type of forensic DNA analysis. As a practical matter, the technique had a serious limitation: RFLP analysis required large amounts of high quality DNA, a condition that was hard to meet with evidence found at many crime scenes. Scientists began to improve forensic DNA analysis by incorporating the polymerase chain reaction (PCR) technique to duplicate a billion copies of a target nucleotide sequence.

Yet the PCR technique presented a new challenge. Traditional RFLP analysis focused on repeated nucleotide sequences in minisatellites, which have sequences of 6 to 100 nucleotides that repeat two to several hundred times. PCR could duplicate DNA segments of about 1,000 to 2,000 nucleotides, a length much shorter than many minisatellite targets. Scientists switched to microsatellites, which have unit lengths that typically peak at 7 nucleotides, repeated 5 to 100 times. Peter Gill, then a researcher at the Forensic Science Service, headquartered in the UK, devised a PCR-based method, which enabled the simultaneous analysis of multiple microsatellite targets called a **short tandem repeat (STR)**. Human chromosomes have thousands of different types of these short repeating sequences, and on each chromosome, the number of repeats varies greatly from person to person. STRs typically have repeating sequences of two to five bases, and the total size of a DNA region containing particular STRs is small, usually fewer than 500 bases. The small size means that only minute amounts of DNA are required for analysis, and that analysis can be performed on partially degraded DNA. Using PCR-STR, technicians can analyze DNA traces found on licked envelopes, cigarette butts, eating utensils, chewing gum, postage stamps, and many other sources.

In 1997, the FBI devised a standard series for DNA profiling that consisted of 13 STRs, each with a sequence of four nucleotides. Why bother analyzing more than one STR? DNA typing results are expressed as probabilities. A DNA profile probability indicates the likelihood that a person chosen at random from a certain population has DNA that fits the DNA profile of the sample obtained from a crime scene. If a lab performed PCR-STR analysis for a single location in one chromosome, there would be a small probability that two random DNA samples would produce the same profile. However, the probability of two random DNA samples producing

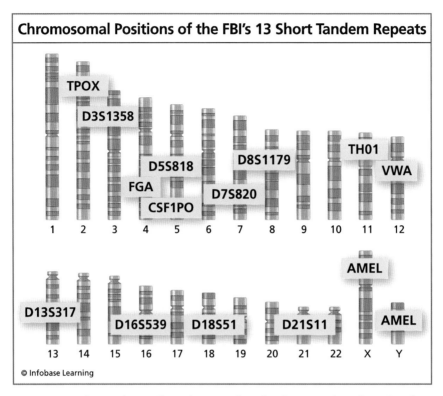

FIGURE 7.2 The FBI has collected 13 core loci for their Combined DNA Index System (CODIS) database.

the same profile becomes minute after combining results of STR analysis from multiple chromosomal locations. Except for identical twins, the likelihood that any two individuals have genomes with the same 13 STR profile may be one in one billion, or lower.

The U.S. forensic science community adopted the FBI's 13 STR method in 1996, and the technique has been nationally accepted as the standard for human DNA identification.

CONTROVERSIES ABOUT FORENSIC DNA ANALYSIS
DNA Profile Databases

The FBI selected the 13 STRs to serve as the genetic markers for its **Combined DNA Index System Program (CODIS)**, an automated DNA information processing and telecommunications system. The CODIS program links DNA databases and transfers data between the databases. Data about DNA profiles are stored in three types of databases, as summarized by this serving of alphabet soup:

$$LDIS \rightarrow SDIS \rightarrow NDIS$$

The FBI structured CODIS as a distributed database in which local authorities generate DNA profiles that flow to state and national levels. At the local level, crime labs operated by police departments, sheriff's offices, or other police agencies create their own databases of forensic DNA profiles. Organizations that are part of this Local DNA Index System (LDIS) can also upload DNA profiles to a state database. Each state participating in the CODIS Program has a single State DNA Index System (SDIS) that contains DNA profiles generated by local labs and DNA profiles created by state labs. SDIS enables state law enforcement agencies to compare DNA profiles that are generated by local labs and state labs. Agencies that are part of the SDIS can upload DNA profiles into the National DNA Index System (NDIS). The FBI also contributes DNA profiles to the NDIS database. NDIS provides a mechanism for crime labs throughout the country to exchange and compare DNA profiles. The federal database stores convicted offender DNA profiles, forensic DNA profiles developed from evidence in criminal investigations, DNA profiles from individuals arrested for certain offenses, and DNA profiles of unidentified human remains, missing persons, and relatives of missing persons. As of September 2010, the NDIS stored more than 8,939,031 offender profiles and 337,988 forensic profiles, the two major categories of profile data.

Here's how CODIS works: After law enforcement specialists recover biological evidence at a crime scene, technicians extract DNA and analyze it to generate a DNA profile, which is then expressed as a series of numbers and entered into the CODIS system. The CODIS software stores the DNA profile data, an identifier of the agency that submitted the profile, and a specimen identification number. The DNA profile is searched against records of convicted offender DNA profiles, and possibly, arrestee DNA profiles. If DNA profile comparisons uncover a 13 STR match, or "hit," then investigators may use the result to obtain a fresh DNA sample from the suspect. The new sample will be used in a confirmation test and, possibly, during a trial.

The DNA profile from biological evidence will also be searched against the forensic index of crime scene DNA profiles. The comparison may provide a link between two crime scenes previously considered

CSI EFFECT: WHAT ARE YOU?

Under the diffused illumination of blue accent lighting, a technician rapidly churns out a DNA analysis. Within minutes, a blood spatter expert analyzes a biological sample to generate a DNA profile, runs the results against database profiles, and gets a result. *CSI: Crime Scene Investigation*, *Dexter*, and other television shows perpetuate myths about forensic science by blurring the line between fact and fiction. The myths have spawned a phenomenon known as the *CSI* effect.

Prosecutors who complain about the *CSI* effect say that shows like *CSI* leave potential jurors with impossibly high expectations about how forensic science analyses routinely solve a case. If the government lacks any forensic evidence to support its case, then prosecutors may present experts—called negative evidence witnesses—to assure jurors that investigators often cannot find DNA and other evidence at crime scenes. Defense attorneys complain that, thanks to the *CSI* effect, jurors place too much faith on scientific results, and they may not appreciate that human or technical error can compromise forensic analyses. Police investigators grumble that a skyrocketing demand by juries to see DNA evidence has increased the amount of evidence submitted to labs already facing hefty backlogs for DNA analysis.

One infamous aspect of the *CSI* effect is that forensics-based shows leave viewers with a feeling that they understand the science and its application in the courtroom. According to one well-known anecdote, jurors in a murder trial became concerned that the blood on the defendant's coat had not been analyzed to generate a DNA profile. The judge explained that the government did not need to perform a DNA analysis; the defendant had admitted to being present at the murder scene.

In their book, *The Real World of a Forensic Scientist* (2009), forensic scientist Dr. Henry C. Lee and his coauthors see another aspect of the *CSI* effect. "Television shows such as *CSI* may well have some influence," they say, "but the effect has been positive as well as negative: As more interest has arisen from the public in forensics, so has funding from Congress and avenues for training investigators have increased."

to be unrelated. Linking crimes in this way can help law enforcement officials in multiple locations to coordinate their investigations and share leads that they have developed independently. A perpetrator may be linked to a number of crime scenes before that person's identity is discovered.

Expansion of DNA Database Laws and Collection Practices

During the late 1980s, Colorado required sex offenders to provide a DNA sample to law enforcement officials. This first U.S. DNA data bank law was enacted with the theory that a person convicted of a sex offense was likely to repeat the crime, and therefore, quick access to a DNA profile could identify a repeat offender. Soon, other states enacted laws that required convicted felons to provide DNA for analysis. Over the years, states expanded the categories of individuals who must donate a sample for a DNA databank. Currently, all states require the collection of DNA from sex offenders, and most states collect DNA from felony offenders. Many states obtain DNA from juveniles convicted of certain crimes and from individuals convicted of certain misdemeanors. Some states also require collection of DNA from probationers and parolees, as well as individuals arrested for certain crimes. Several states allow law enforcement agencies to collect DNA from individuals identified as suspects in an investigation and retain the DNA profile for several years.

Congress has also expanded the collection of DNA by federal law enforcement officials:

- The DNA Analysis Backlog Elimination Act of 2000 authorized DNA collection from individuals in federal custody and people convicted of certain violent crimes who are probationers, parolees, or on supervised release from prison.
- The 2001 U.S.A. PATRIOT Act expanded the types of qualifying federal offenses that require DNA collection.
- The Justice for All Act of 2004 authorized DNA collection from all persons convicted of felonies under federal law, and allowed state law enforcement agencies to upload DNA profiles to CODIS from anyone convicted of a crime.

- The DNA Fingerprint Act of 2005 authorized federal officials to collect DNA samples from individuals who are arrested and from non-U.S. citizens detained under federal authority.

Prison inmates challenged state DNA databank laws by alleging that the laws violated the U.S. Constitution's Fourth Amendment rights against unreasonable searches and seizures. However, courts upheld state DNA collection statutes on the grounds that an inmate's limited privacy interests are outweighed by the government's interest in obtaining reliable DNA identification evidence for possible use in solving crimes. Typically, federal courts have approved laws that allow the collection of biological samples from prison inmates convicted of a qualifying federal offense and as a condition of supervised release from prison. This is not a settled area of law, however, because judges differ in their opinions. One federal court, for example, decided that forced removal of blood from parolees conflicts with rights guaranteed by the U.S. Constitution.

Retaining DNA profile data about arrestees who are not convicted for a crime remains a very controversial aspect of DNA databank laws. In 2006, the Minnesota court of appeals ruled that an individual's privacy interests trumped the state's interest in analyzing DNA. Those parts of the state's DNA arrestee law, the court decided, "that direct law-enforcement personnel to take a biological specimen from a person who has been charged but not convicted violate the Fourth Amendment to the United States Constitution and Article I, Section 10 of the Minnesota Constitution." In 2009, a U.S. district court judge decided that provisions of the DNA Analysis Backlog Elimination Act of 2000, which allow federal law enforcement officials to take a DNA sample from a charged defendant without a warrant, are invalid under the Fourth Amendment. Other courts have approved arrestee DNA collection laws, and supporters of the laws assert that they increase the efficiency of criminal investigations and save lives. The debate about taking biological samples from arrestees continues.

Another method of obtaining a DNA sample also incites protests: surreptitious DNA collection. Depending upon one's point of view, surreptitious can mean secret or sneaky. Police detectives in television shows often rely upon the practice by collecting a suspect's discarded coffee cup, chewing gum, or cigarette butt to extract a DNA sample. Law enforcement agencies argue that surreptitious DNA collection is justified, because an individual has no privacy interest in an abandoned DNA sample.

Debates About Familial Searching

Initially, the FBI would not release the results of a CODIS search unless it yielded a complete, 13 STR match. In 2006, the FBI announced a temporary plan to release partial matches detected with CODIS. Law enforcement agencies use partial matches for "familial searching," a method to identify possible relatives of a suspect in a criminal investigation. Law enforcement officials and prosecuting attorneys argue that familial searching generates new leads, which can result in arrests and convictions. Privacy advocates

COURT CONDONES COLLECTION CON

One famous case about surreptitious DNA collection began with a Seattle police investigation of the 1982 murder of a 13-year-old girl. A teenage boy named Athan was one suspect, but he was not charged due to a lack of evidence. Twenty years later, the Seattle Police Department's cold case detectives sent stored biological evidence from the crime scene to the police lab for DNA analysis. Although the lab generated a DNA profile, it failed to match profiles in state and federal databases. The detectives wanted a DNA sample from Athan, who now lived in New Jersey. So, they sent him a letter from a fictitious law firm, inviting Athan to join a fabricated class action lawsuit. Athan signed the agreement and mailed it. The detectives forwarded the unopened letter to the lab where a technician generated a DNA profile from saliva on the envelope flap. Athan's DNA profile matched the profile from the 1982 crime scene. Detectives arrested Athan, obtained a warrant for another DNA sample, and confirmed a match between the crime scene DNA and Athan's DNA. Athan was found guilty of second degree murder. He then appealed the conviction.

In 2007, the Washington State Supreme Court deemed the DNA evidence admissible under both the U.S. Constitution and the state constitution. "No recognized privacy interest exists in voluntarily discarded saliva," the court ruled, "and a legitimate government purpose in collecting a suspect's discarded DNA exists for identification purposes." The court asserted that, although the detectives' ruse violated certain statutes, the deception was not so outrageous or shocking to warrant a dismissal of the case.

protest that familial searching places innocent people under surveillance. Several states have laws that authorize familial searching, and law enforcement agencies in the UK have used the technique for years.

One concern about familial searching is that a partial match may not identify a genetic relative. Two DNA profiles can partially match statistically even though the profiles were generated from DNA samples of individuals who are not related. Despite apprehensions about familial searching, the practice may become more common after the success in the Grim Sleeper case. When a 20-year investigation of a serial killer called the Grim Sleeper failed to generate useful leads, the Los Angeles Police Department (LAPD) requested a familial search with DNA profiles in state databases. The search produced 200 DNA profiles of people who might be related to the serial killer. One DNA profile shared genetic markers with DNA recovered at 15 crime scenes of the Grim Sleeper. Further rounds of analyses led police to believe that they had identified the suspected serial killer's son, whose DNA profile had recently been added to the database following a felony weapons conviction. The investigation led to an arrest of the alleged serial killer in July 2010.

The successful outcome inspired comments with opposing viewpoints: "This will change the face of policing in the United States," said LAPD Chief Charlie Beck at the press conference after the arrest. James Alan Fox, professor of criminology, law, and public policy at Boston's Northeastern University, had a different take on the subject: "Laws permit the collection and storage of DNA data on certain convicted offenders for use in potentially linking future crimes to these same criminals, but not to their blood relatives who happen to have similar genetic profiles," Fox posted on *The Boston Globe* Web site. Fox voiced concern that the investigation of innocent blood relatives may be a first step in an expanding invasion of privacy.

EVERYONE INTO THE DNA DATABASE?

While debates about the government collection of DNA samples from adults continue, some commentators suggest an even more radical idea: Government agencies should take a DNA sample from everyone at birth to create a population-wide forensic database. Currently, the U.S. government cannot afford the costs of such an expansive change in forensic DNA analysis and DNA profile data storage. Yet advances in technology may create the opportunity to implement such a program in a financially reasonable way. At that time, society would have to decide if benefits to the public outweigh the intrusion into privacy.

Evolution and Intelligent Design

In the biological sciences, **evolution** refers to gradual changes that occur in plants, animals, and other life forms over generations. Long before the discovery of genes and DNA, scientists investigated evolution by comparing structures in cells and by comparing physical traits of plants and animals. Today, scientists know that differences in cell structure and physical traits are due to genetic mutations. A mutation of a gene can alter cellular functions, and an altered cellular function may offer an advantage for an organism. If so, then the organism should be able to pass on a copy of the mutated gene to offspring. As generations pass, an increasing number of organisms in a population should have the mutated gene, and as a result, members of the population may differ physically from their ancestors.

For more than 150 years, the subject of evolution has been a lightning rod for controversy. During the early twentieth century, people viewed evolution as an idea that contradicted their religious beliefs, and they tried to ban the teaching of evolution from public schools. More recently, opposition to the teaching of evolution has taken the form of demands that schools teach creationism or intelligent design as an alternative explanation for the variety of life on Earth.

THE THEORY OF EVOLUTION

In 1831, 22-year-old Charles Darwin, who had an interest in science and natural history, accepted an offer to join a survey ship for a two-year trip around the world. Darwin boarded the 90-foot-long (27-meter-long) brig, HMS *Beagle*, which set sail from Plymouth, England, under clear skies on December 27, 1831. Many discoveries awaited the young scientist. The first was that he became seasick even when sailing in calm weather. Another surprise awaiting him was that the trip would take five years to make, not two.

Darwin took every opportunity to investigate dry land. While exploring South America, he came upon evidence of upheavals in the land. While trekking through the Andes Mountains, for example, Darwin found petrified trees resembling trees he had seen at sea level, 7,000 feet (2 kilometers) below. One popular idea at the time was that the Earth was only 6,000 years old and unchanging, and that plant and animal species remained the same since the dawn of creation. Darwin's explorations led him to accept a recent theory that the Earth's features had been slowly changing over great periods of geological time.

Much to the disapproval of *Beagle* Captain Robert FitzRoy, Darwin collected crates of fossils. Some of these fossils represented unknown species, while others bore some similarities to living animals. In different parts of the world, Darwin also saw animals belonging to two different species that nevertheless closely resembled each other. This led him to wonder if they had shared a common ancestor.

In the fall of 1835, the *Beagle* arrived at the islands of Galápagos, where Darwin observed two types of iguanas. One type of iguana could stay underwater for at least an hour, and swam with a flattened tail. Other iguanas lived on land and had limbs and strong claws adapted for crawling over jagged terrain. Both types of animals were well-suited for their environments. Darwin also discovered plants and animals that seemed related to species he had seen on the mainland, but these newly observed species had unique physical variations. For example, many of the island animals had coloration that blended with the surrounding lava fields. He wondered if species of plants and animals from the mainland had reached the islands and then eventually adapted to the new environment through many generations.

After Darwin returned to England, he organized his collections and observations about the plants and animals he had seen, both living and fossilized. Darwin theorized that plants and animals must be slowly and constantly changing. In the competition for food or land, an organism with a different trait that provided a competitive edge would be more likely to leave offspring than those that lacked the advantage, he concluded. As the descendants of those with an advantageous trait reproduced, and their descendants multiplied, more members of the species would have the useful trait. The species would alter through this natural selection.

By 1842, Darwin had formulated the basis for his book, *The Origin of Species*. Yet he delayed publication because he knew that his ideas contradicted a widely held view of humans as above and outside of nature. His theory indicated that humans and apes had evolved from a common ancestor. In 1858, Darwin learned about British naturalist Alfred Russel Wallace's theory of evolution by natural selection. After decades of labor, Darwin did not want to be seen as merely following in Wallace's footsteps, so he finished his book and published it the following year.

Scientists have identified five theories that make up Darwin's idea about evolution, a process that Darwin called descent with modification through natural selection.

- Living things constantly change through generations.
- All life descended from a common ancestor. Darwin represented this idea as an evolutionary tree with ancient life forms at the bottom and descendants branching off along the trunk. An evolutionary tree contradicted a popular view that species are not related to each other and have remained unaltered since their creation.
- New species are produced by isolating and transforming populations of old species.
- Large differences in physical traits appear over long periods of time by an accumulation of smaller changes.
- Natural selection is one engine for evolutionary change. Adaptation to an environment is the result of natural selection.

Scientists have refined the theory of evolution in light of advances in genetics, embryology, paleontology, molecular biology, and other disciplines. The development of the field of genetics is one of the most important advances since Darwin's time. Scientists now understand that variations in genes result in inheritable variations of physical traits. Today, biological evolution is often divided into microevolution and macroevolution. Microevolution refers to changes in the occurrence of different forms of genes within a population. Macroevolution refers to large alterations in living things, including the appearance of new life forms, such as mammals, and mass extinction of species. While scientists continue to

SURVIVAL OF THE FITTEST CAN CREATE UNFIT TRAITS

Hiccup sufferers can blame evolution says Neil Shubin, a professor at the University of Chicago. A bout of hiccups can result from interference with nerves that control breathing, nerves inherited from fish that had their gills, a fish's breathing apparatus, closer to the neck. In humans, these nerves follow a long, winding course from the base of the skull, past the neck, and through the chest cavity. Hiccups can develop when these lengthy, twisting nerves become irritated. The hic of a hiccup may be an action handed down from tadpoles, which breathe with lungs or gills. When a tadpole wants to breathe with its gills, it closes the passage into the lungs and forcefully pumps water into its mouth. A hiccup's hic—the quick intake of breath and block of the throat—mimics this action.

Health problems more serious than hiccups may be blamed on evolutionary trade-offs. As one example, gene variations that protect against malaria first emerged in Africa. However, these altered genes can also produce sickle cell disease. Certain genes help the body to retain salt and prevent dehydration, a useful trait in tropical areas of the world. Yet these same genes increase the risk for high blood pressure.

According to the hygiene hypothesis, the human immune system evolved in a world that required a constant battle against bacteria, parasites, and viruses. Thanks to public health reforms, people in many parts of the world live in a relatively pathogen-free environment. The human immune system, geared up for battle, needs to find something to attack. So, sometimes the immune system overreacts and attacks the body. This results in an increased occurrence of allergies, asthma, and other autoimmune diseases.

The hygiene hypothesis indicates that the health disorders that humans face today occur because people have altered their environment. What will happen as humans continue to transform civilization? Yale evolutionary biologist Stephen Stearns told *The Wall Street Journal* that the gap between human traits and modern culture may widen in the future. "My students keep asking me, 'Are we adapting to the computer age?'" Stearns said. "But technology and culture are changing so fast that our genes can't keep up with it."

modify ideas about evolution, one thing does not change: Many people object to the teaching of evolution on religious grounds.

TEACHING EVOLUTION: BATTLE FOR SURVIVAL IN THE CLASSROOM

Readers of Charles Darwin's *Origin of Species* (1859) discovered a view of evolution driven by random variations in physical traits and a callous survival of the fittest selection mechanism. These ideas offended many Christians who believed in a structured world that reflected an intelligent design by a loving Creator. Advocates of Darwinism and

FIGURE 8.1 Science professor Maude Stout points to blackboard on which appear quotes from the Bible (*Genesis* 1:25) and from Charles Darwin's theory of evolution as she teaches creationism at Bob Jones University in Greenville, South Carolina, in 1948.

religious leaders debated. By the turn of the twentieth century, however, some religious leaders promoted a philosophy that evolution served God's purpose. In England and the United States, many leading Protestant theologians simply decided that theology would confine itself to areas of faith and morality, while science would focus on the natural world.

However, the Protestant fundamentalist movement opposed liberal efforts to reconcile faith and science. Following World War I, many religious leaders decided that Darwinism eroded the nation's moral foundations by substituting materialism for faith in the Bible. During the 1920s, an American crusade against teaching evolution in schools gained momentum. Two factors fueled a sense of urgency for this protest: a Darwinian view of human evolution began to appear in textbooks being published at the time and a striking rise in school attendance. Both of these developments incited fears that Darwin's ideas might affect schoolchildren's faith.

William Jennings Bryan, a three-time Democratic candidate for president, became a leader of the anti-evolution movement. Bryan believed that people used the idea of natural selection to justify a killing of the weak by the strong. Bryan found support for his fears in the American eugenics movement, which called for programs to imprison or sterilize people deemed unfit for the struggle of existence.

Court Battles Over the Teaching of Evolution

During the 1920s, 20 state legislatures considered bills to prevent the teaching of evolution in public schools. On March 23, 1925, Tennessee Governor Austin Peay signed a bill into law that created a misdemeanor punishable by fine for any Tennessee public school teacher "to teach any theory that denies the story of the Divine Creation of man as taught in the Bible, and to teach instead that man has descended from a lower order of animals." Peay might have thought that nobody would enforce a law that was intended to reassure voters that their legislators still believed in Genesis.

Deciding that Tennessee's anti-evolution statute threatened academic freedom, the New York City-based American Civil Liberties Union published an offer for a Tennessee teacher who would be willing to test the law in court. Meanwhile, business leaders in Dayton, Tennessee, saw this offer as an opportunity to attract tourism dollars with a prominent trial. Their enthusiasm gave birth to what became known as the "Drugstore

FIGURE 8.2 During the Scopes Monkey Trial in 1925, defense lawyer Clarence Darrow rests on a desk in the crowded Rhea County Courthouse in Dayton, Tennessee. Scopes, a high school biology teacher, was on trial for violating the Butler Act, a Tennessee law that forbade the teaching of the theory of evolution in public schools because it contradicts the Bible. Scopes was convicted and fined $100. Although the state supreme court later overturned the decision, the Butler Act remained in the books until 1967.

Conspiracy." During a meeting in Frank E. Robinson's drugstore, Dayton city attorneys agreed to prosecute a violation of the new law. They contacted John T. Scopes, a 24-year-old high school football coach and occasional substitute biology teacher. While Scopes admitted that he had used a state-approved textbook that presented Darwin's theories, he could not recall if he had taught evolution. However, the Drugstore Conspirators were not concerned about this detail, and Scopes agreed to be the defendant in a test case of Tennessee's anti-evolution law. Soon, a justice of the peace swore out a warrant for Scopes' arrest.

The trial took place during the summer of 1925. Famed attorney Clarence S. Darrow headed the Scopes defense team, while the special prosecutor was none other than William Jennings Bryan. At first, the defense wanted to prove that the anti-evolution law was invalid. However, the judge did not want to hear any objections to the statute. Then, the defense team tried to convince the jury that evolution did not contradict the Bible's account of Genesis. This also turned out to be the wrong strategy. After the prosecution's witnesses testified that Scopes had taught the theory of evolution, Darrow asked the judge to instruct the jury to bring a guilty verdict, so that the defense team could pursue arguments about the statute's constitutionality on appeal. Bryan graciously offered to pay

FIGURE 8.3 Biology teacher Susan Epperson, who challenged Arkansas's ban on the teaching of the theory of evolution, is shown at her desk at Little Rock Central High School in Little Rock, Arkansas on August 13, 1966.

Scopes' $100 fine. The Tennessee Supreme Court thwarted the defense's plans to take the case to the U.S. Supreme Court by overturning the conviction on a technicality.

Tennessee's ban on teaching evolution persisted for decades. As other state and local governments frowned upon evolution, the subject faded from new textbooks and teachers became reluctant to mention the subject. The topic of evolution enjoyed new life during the 1960s. Concerned that the Soviet Union would surpass the United States in science, the U.S. government funded science education programs. By now, studies in genetics and paleontology supported many of Darwin's ideas. The revival of evolution as a school subject also revived battles between evolutionists and anti-evolutionists.

In *Epperson v. Arkansas* (1968), the U.S. Supreme Court struck down Arkansas's 1928 anti-evolution law as a violation of the First Amendment doctrine of church and state separation. "Government in our democracy, state and national, must be neutral in matters of religious theory, doctrine, and practice," wrote Justice Abe Fortis. "It may not be hostile to any religion or to the advocacy of no-religion. . . . The First Amendment mandates governmental neutrality between religion and religion, and between religion and nonreligion."

Since the *Epperson* decision clearly stated that the Supreme Court demanded religious neutrality from the government, anti-evolutionists responded with a different tactic during the 1970s. They portrayed creationism as "creation science" and demanded equal time in the classroom for their alternative to evolution. By the 1980s, 27 states introduced legislation requiring equal time for creation science. Several states enacted equal time laws. However, judges recognized that creation science was fundamentalist religion in disguise.

In *Edwards v. Aguillard* (1987), the U.S. Supreme Court struck down a Louisiana equal time law. "The Louisiana Creationism Act advances a religious doctrine by requiring either the banishment of the theory of evolution from public school classrooms or the presentation of a religious viewpoint that rejects evolution in its entirety," wrote Justice William J. Brennan, Jr. "The Act violates the Establishment Clause of the First Amendment because it seeks to employ the symbolic and financial support of government to achieve a religious purpose."

A new strategy evolved: Promotion of "intelligent design," which advances the idea that an Intelligent Designer must be responsible for at least some physical and functional aspects of living organisms, because

EVOLUTIONARY HITCHHIKERS

A basic idea of evolution is that populations of living things adapt to a changing environment through the selection of traits that help them to survive and reproduce. For example, an animal that has a gene for a favorable trait will have an improved chance to live long enough to pass on the gene to offspring.

A research team led by University of Rochester biologist John Jaenike discovered that evolution can take a shortcut. The scientists studied *Drosophila neotestacea*, a species of North American fly that has been under attack by parasitic roundworms. Female roundworms infest young female flies and become engorged with thousands of eggs. After the roundworm's eggs hatch, they spread throughout the fly's body. The fly becomes so weak that it cannot reproduce. Although the flies lack a natural, effective defense against the roundworms, a type of bacteria called *Spiroplasma* can help. If a female fly is infested with the bacteria, invading roundworms become sick and produce few offspring, which do not interfere with the host fly's ability to reproduce. The flies pass the bacteria to their own offspring, so that they have a ready defense against the parasitic roundworms.

Jaenike and his team found that *D. neotestacea* flies had been heavily infested with the roundworms during the 1980s. Examination of preserved flies revealed that about 10% of flies living in the eastern United States carried *Spiroplasma* bacteria. By 2008, about 80% of tested flies harbored the bacteria. The selective pressure of the invading roundworm favored *Spiroplasma* infection, which flies pass from one generation to the next. Even though the flies lacked a gene to fight the roundworm, they adapted by taking on the bacteria hitchhiker.

these traits are too complex to be the result of natural processes. Pennsylvania's Dover Area School District devised a policy that required teachers to read a statement about intelligent design to students in the ninth grade biology class. "Because Darwin's Theory is a theory," teachers would inform students, "it continues to be tested as new evidence is discovered. The Theory is not a fact. Gaps in the Theory exist for which there is no evidence.... Intelligent Design is an explanation of the origin of life that differs from Darwin's view."

A group of parents sued in a federal district court, alleging that the school policy was unconstitutional. During the six-week trial of *Kitzmiller v. Dover Area School District* (2005), Judge John E. Jones II heard testimony from scientists, intelligent design advocates, parents, teachers, school board members, and others. At the end, the judge decided that the policy was an attempt to impose a religious view of biological origins into a biology course, which violated the U.S. Constitution's mandate to separate church and state.

OPPOSITION TO THE TEACHING OF EVOLUTION CONTINUES TO ADAPT

In his *Kitzmiller* decision, Judge Jones noted that, "Repeatedly in this trial, Plaintiffs' scientific experts testified that the theory of evolution represents good science, is overwhelmingly accepted by the scientific community, and that it in no way conflicts with, nor does it deny, the existence of a divine creator." Yet opposition to teaching evolution in public schools endures.

During an interview conducted for the Public Broadcasting Service's *NOVA* program, Dr. Kenneth Miller, who testified at the Dover Area School District trial as an expert witness for the plaintiffs, gave his opinion about why evolution is so controversial.

> Evolution concerns who we are and how we got here. And to an awful lot of people, the story of evolution, the story of our continuity with every other living thing on this planet, that's not a story they want to hear.
>
> They favor an entirely different story, in which our ancestry is separate, our biology distinct, and the whole notion of our lineage traceable not to other organisms, but to some sort of divine power and divine presence. But it's absolutely true that our ancestry traces itself along the same thread as that of every other living organism. That, for many people, is the unwelcome message, and I think that's why evolution has been, is, and will remain such a controversial idea for many years to come.

Anti-evolutionists have tried to ban the teaching of evolution by law and enact laws that demand equal time with creation science or intelligent design. Recent tactics in the struggle against the teaching of evolution include promotion of the idea that not only is evolution controversial among scientists, but it is also a "theory in crisis."

Glossary

adult stem cells Multipotent stem cells that reside in small numbers in the organs and tissues of adults

amino acid The chemical building block of a protein

antibody A protein that binds with a substance foreign to a body

antigen A substance that the immune system recognizes as foreign

base A molecule that forms part of DNA and RNA

base pair Two bases from two nucleotides, held together by weak bonds, in a double-stranded DNA or RNA molecule

cell culture Cells grown in a laboratory

chromosome A structure in a cell that contains DNA

cleavage site A target nucleotide sequence for a restriction enzyme

clone An animal that is a copy of another animal

codon A group of three nucleotides. Most codons code for an amino acid.

Combined DNA Index System Program (CODIS) FBI's software program that enables local, state, and federal crime labs to compare and exchange DNA profiles

cord blood stem cells Stem cells in the blood of an umbilical cord

covalent bond A chemical link created when two atoms of different chemicals share electrons

deoxyribonucleic acid (DNA) A nucleic acid molecule that encodes genetic information and contains deoxyribose sugar

deoxyribose A five-carbon sugar called ribose that is missing an oxygen atom (deoxy-) on its second carbon and is part of a nucleotide that makes up DNA molecules

differentiate Process by which an unspecialized cell develops into a cell with specialized structures and functions

embryonic stem cell line   Group of embryonic stem cells that proliferates in culture for months or years, remains undifferentiated, and retains pluripotent capacity

enzyme   A protein that increases the rate of a chemical reaction

epigenetic   Changes in a gene's function that do not involve altering nucleotide sequences in DNA

eugenics   An effort intended to improve the human race, which would decrease genetic diversity

evolution   In the context of biology, the term refers to gradual changes that occur in plants, animals, and other life forms over generations

ex vivo   Experiment performed on living cells or tissues after removal from an organism

gene   A nucleotide sequence that encodes a protein or a functional RNA molecule

gene expression   Process in which information stored in a DNA molecule is used to make a product, such as a protein

gene therapy   Treatment of a genetic disorder that aims to replace or to supplement an abnormal gene with a normal gene

genetic code   The collection of 64 codons that specify 20 amino acids and the signals for stopping protein synthesis

genetic disease   A disorder resulting from DNA mutations that create abnormal nucleotide sequences in a person's genome

genetic test   Analysis of a tissue sample to examine the function of a gene or the presence of a gene mutation

genome   The complete set of an organism's genes within the collection of chromosomes found in the cell nuclei

genomic imprinting   An epigenetic effect in which the parental source of a gene affects gene activity

germline genetic modification   A theoretical treatment for a genetic disease in which one or more transgenes is introduced into an egg cell or sperm cell with the objective of benefitting future generations

hematopoietic stem cells   Multipotent stem cells that differentiate into blood cells

induced pluripotent stem cell   Stem cell produced by introducing genes into a differentiated cell

informed consent   A rule that requires that a potential human subject of an experiment must be educated about the risks of the experiment, must understand the risks, must give consent voluntarily, and must be mentally capable of giving consent

in vivo   An experiment or procedure performed in the living body of an organism

messenger RNA (mRNA)   An RNA molecule that transmits genetic information from DNA to a cell's protein-making apparatus

mitochondria   Intracellular structures that function as a cell's power plant

multipotent   Ability of a cell to differentiate into a limited number of cell types

mutation   A change in the nucleotide sequence of a DNA molecule or a change in the amino acid sequence of a protein

nuclear transfer   Swapping one nucleus for another to produce a clone

nucleic acid   A molecule of DNA or RNA

nucleotide   The monomer (a simple molecule that combines with similar or identical molecules to form a polymer) of DNA which contains a sugar molecule, a chemical group that contains phosphorus, and a base

nucleus   Intracellular structure that contains most of a cell's DNA

organelles   Membrane-bound structures that perform functions within a cell

plasmid   A ring of DNA found mostly in bacteria and capable of independent replication

pluripotent   Ability of a cell to develop into many types of specialized, differentiated cells

polymer   A large chemical made by combining smaller chemical units

polymerase chain reaction (PCR)   A process that produces millions of copies of a short nucleotide sequence

protein   A polymer of amino acids

protein synthesis   Production of a protein polymer formed by the addition of amino acids that connect with each other by covalent bonds

Glossary

recombinant DNA   DNA that has been altered in the lab by the addition or deletion of nucleotide sequences

reproductive cloning   Nuclear transfer of an egg cell nucleus for a somatic cell nucleus with the goal of producing an embryo that will be implanted in a surrogate mother and allowed to mature to an infant

restriction enzyme   A protein that cleaves a DNA molecule at or near a certain nucleotide sequence

restriction fragment length polymorphism (RFLP)   Variation between individuals in the pattern of DNA fragments obtained by treating DNA with a restriction enzyme

ribonucleic acid (RNA)   A nucleic acid molecule that can encode genetic information and contains ribose sugar

ribose   A five-carbon sugar that forms part of a nucleotide that makes up RNA molecules

short tandem repeat (STR)   Short nucleotide sequences in DNA that are repeated many times

somatic cell   A cell other than an egg cell or a sperm cell

stem cell   An undifferentiated cell that divides to produce daughter cells that either remain stem cells or differentiate into specialized cells

terminally differentiated   The condition of a specialized cell that does not develop into other types of cells

therapeutic cloning   Nuclear transfer of an egg cell nucleus for a somatic cell nucleus with the goal of producing an embryo that will be used to develop an embryonic stem cell line to treat a disease

totipotent   The ability of a cell to develop into a whole organism or into cells of any of its tissues

transcription   The process of making an RNA copy of a DNA nucleotide sequence

transgene   A gene that is transferred to a cell, for example, in gene therapy

transgenic   Describes an organism that has been genetically altered using recombinant DNA technology

translation   The process of making a protein using genetic code information in messenger RNA

utility patent A limited property right granted by the U.S. government for a new and useful invention

vector In biotechnology, a DNA molecule that can be used to deliver a transgene

Bibliography

Abrahams, Edward and Mike Silver. "The Case for Personalized Medicine." *Journal of Diabetes Science and Technology* 3 (2009): 680–684.

Aiuti, Alessandro, Federica Cattaneo, Stefania Galimberti, Ulrike Benninghoff, Barbara Cassani, et al. "Gene Therapy for Immunodeficiency Due to Adenosine Deaminase Deficiency." *New England Journal of Medicine* 360 (2009): 447–458.

Alberts, Bruce, Dennis Bray, Karen Hopkin, Alexander Johnson, Julian Lewis et al. *Essential Cell Biology,* 3rd Edition. New York: Garland Science, 2009.

Aronson, Jay D. *Genetic Witness: Science, Law, and Controversy in the Making of DNA Profiling.* New Brunswick: Rutgers University Press, 2007.

Association for Molecular Pathology et al. v. United States Patent and Trademark Office et al., Docket No. 09 Civ. 4515 (March 29, 2010). Available online. URL: www.aclu.org. Accessed February 10, 2011.

"At-Home Genetic Tests: A Healthy Dose of Skepticism May Be the Best Prescription." Federal Trade Commission Web site, July 2006. Available online. URL: http://www.ftc.gov. Accessed February 10, 2011.

Battey, James F., Jr. and Laura K. Cole. "A Stem Cell Primer." *Pediatric Research* 59 (2006): 1R–3R.

Beck, Melinda. "Obesity? Big Feet? Blame Darwin." *The Wall Street Journal*, February 23, 2010.

Behfar Atta, Satsuki Yamada, Ruben Crespo-Diaz, Jonathan J. Nesbitt, Lois A. Rowe et al. "Guided Cardiopoiesis Enhances Therapeutic Benefit of Bone Marrow Human Mesenchymal Stem Cells in Chronic Myocardial Infarction." *Journal of the American College of Cardiology* 56 (2010): 721–734.

Bellomo, Michael. *The Stem Cell Divide.* New York: AMACON, 2006.

Besser, Richard, Susan Schwartz, and Lara Salahi. "Stem Cell Cornea Fix: Better Vision May Be in Your Own Eyes." ABC News Web site, June 23, 2010. Available online. URL: http://abcnews.go.com. Accessed February 10, 2011.

Bestor, Timothy H. "Gene Silencing as a Threat to the Success of Gene Therapy." *Journal of Clinical Investigation* 105 (2000): 409–411.

"Bioethics and Patent Law: The Cases of Moore and the Hagahai People." *WIPO Magazine*, September 2006. Available online. URL: http://www.wipo.int/wipo_magazine/en/2006/05/article_0008.html. Accessed February 10, 2011.

Bischoff, Steve R., Shengdar Tsai, Nicholas Hardison, Alison A. Motsinger-Reif, Brad A. Freking, et al. "Characterization of Conserved and Nonconserved Imprinted Genes in Swine." *Biology of Reproduction* 81 (2009): 906–920.

Blank, Alan. "Alternative Evolution: Why Change Your Own Genes When You Can Borrow Someone Else's?" University of Rochester Web site, July 8, 2010. Available online. URL: http://www.rochester.edu/news. Accessed February 10, 2011.

Branch, Glenn and Eugenie C. Scott. "The Latest Face of Creationism." *Scientific American* 300 (January 2009): 92–99.

Browne, Janet. *Charles Darwin: Voyaging*. Princeton: Princeton University Press, 1995.

Buckelew, Karen. "New UM School of Medicine Study Finds More Variation in Human Genome than Expected." University of Maryland Office of External Affairs Web site, June 24, 2010. Available online. URL: http://www.oea.umaryland.edu. Accessed February 10, 2011.

Campbell, Neil A., Jane B. Reece, Lisa A. Urry, Michael L. Cain, Steven A. Wasserman, Peter V. Minorsky, and Robert B. Jackson. *Biology*, 8th Edition. San Francisco: Pearson/Benjamin Cummings, 2008.

Cass, James. "The Cost of Making Crime Not Pay: Obama, CODIS and Forensic DNA." Genomics Law Report Web site, March 23, 2010. Available online. URL: http://www.genomicslawreport.com. Accessed February 10, 2011.

The Center for Bioethics and Human Dignity. "Position Statement." The Center for Bioethics and Human Dignity Web site, 1999. Available online. URL: http://www.cbhd.org.

Cetrulo, Curtis, L., Kyle J. Cetrulo, and Curtis L. Cetrulo, Jr., eds. *Perinatal Stem Cells*. Hoboken, New Jersey: John Wiley & Sons, Inc., 2009.

Charo, Alta R. "Body of Research—Ownership and Use of Human Tissue." *New England Journal of Medicine* 355 (2006): 1517–1519.

Choi, Charles Q. "Old MacDonald's Pharm." *Scientific American* 295 (September 2006): 24.

Cock, Matthew. "Biopiracy Rules Should Not Block Biological Control." *Nature* 467 (2010): 369.

Committee on Defining Science-Based Concerns Associated with Products of Animal Biotechnology and the Committee on Agricultural Biotechnology, Health, and the Environment, National Research Council. *Animal Biotechnology:*

Science Based Concerns. Washington, D.C.: The National Academies Press, 2002.

Cook-Deegan, Robert. "Gene Patents." In *From Birth to Death and Bench to Clinic: The Hastings Center Bioethics Briefing Book for Journalists, Policymakers, and Campaigns*. Mary Crowley, ed., 69–72. Garrison, NY: The Hastings Center, 2008.

Cookson, Clive. "Mother of All Cells." In *The Future of Stem Cells*. Lionel Barber and John Rennie, eds., A6–A10. New York: Scientific American and Financial Times, 2005.

Cooper, Geoffrey M. and Robert E. Hausman. *The Cell: A Molecular Approach*. 5th Edition. Washington, D.C.: ASM Press, 2009.

Darwin, Charles. *Voyage of the Beagle*. London: Penguin Books, 1989.

Devlin, Thomas M., ed. *Textbook of Biochemistry with Clinical Correlations*. 7th Edition. New York: John Wiley & Sons, Inc., 2011.

Dolan, Maura. "In Grim Sleeper Case, A New Tack in DNA Searching." *Los Angeles Times*, July 10, 2010.

Drummond, Katie. "Darpa's Lab-Grown Blood Starts Pumping." *Wired*, July 9, 2010. Available online. URL: http://www.wired.com. Accessed February 10, 2011.

Eiseman, Elisa and Susanne B. Haga. *Handbook of Human Tissue Sources*. Rand Corporation Web site, 1999. Available online. URL: http://www.rand.org. Accessed February 10, 2011.

Epperson v. Arkansas, 393 U.S. 97 (1968).

Erickson, Britt E. "Lab-Developed Tests Come Under Fire." *Chemical & Engineering News* 88 (August 9, 2010): 24–25.

Evans, Paula C. "Patent Rights in Biological Material." IPFrontline Web site, October 3, 2006. Available online. URL: http://www.ipfrontline.com. Accessed February 10, 2011.

"First Evidence of Genetically Modified Plants in the Wild, Scientists Report." Science Daily Web site, August 6, 2010. Available online. URL: http://www.sciencedaily.com. Accessed February 10, 2011.

Foreman, Carol Tucker. "Killing the 'Frankenfood' Monster: How People Can Love, not Fear, Biotech Food." *The American Enterprise* 15 (March 2004): 30–32.

Fox, James Alan. "Catching the Grim Sleeper." *The Boston Globe* Web site, July 8, 2010. Available online. URL: http://boston.com. Accessed February 10, 2011.

Frieden, Thomas R. and Francis S. Collins. "Intentional Infection of Vulnerable Populations in 1946-1948: Another Tragic History Lesson." *JAMA* Web site,

October 11, 2010. Available online. URL: http://jama.ama-assn.org. Accessed February 10, 2011.

"Gene Testing." U.S. Department of Energy Genome Programs Web site, September 17, 2010. Available online. URL: http://www.ornl.gov/sci/techresources/Human_Genome/medicine/genetest.shtml. Accessed February 10, 2011.

Greger, Michael. "Transgenesis in Animal Agriculture: Addressing Animal Health and Welfare Concerns." *Journal of Agricultural and Environmental Ethics*, 2010. Available online. URL: http://www.springerlink.com.

Griffiths, Anthony J.F., Susan R. Wessler, Richard C. Lewontin, and Sean B. Carroll. *Introduction to Genetic Analysis*, 9th Edition, New York: W.H. Freeman and Company, 2008.

Hanna, Kathi E. "Reproductive Genetic Testing." National Human Genome Research Institute Web site, January 2006. Available online. URL: http://www.genome.gov/10004766. Accessed February 10, 2011.

Hanna, Kathi E. "Germline Gene Transfer." National Human Genome Research Institute Web site, March 2006. Available online. URL: http://www.genome.gov/10004764. Accessed February 10, 2011.

Hanna, Kathi E. "Cloning/Embryonic Stem Cells." National Human Genome Research Institute Web site, April 2006. Available online. URL: http://www.genome.gov/10004765. Accessed February 10, 2011.

Harris, J. "Clones, Genes, and Reproductive Autonomy: The Ethics of Human Cloning." *Annals of the New York Academy of Sciences* 913 (September 2000): 209–217.

Hartwell, Leland H., Leroy Hood, Michael L. Goldberg, Ann E. Reynolds, Lee M. Silver, and Ruth C. Veres. *Genetics: From Genes to Genomes*, 3rd Edition. New York: McGraw-Hill, 2008.

Hatch, Orrin G. "Testimony of the Honorable Orrin G. Hatch Before the House Senate Subcommittee on Criminal Justice—Stem Cell Research." United States Senator Orrin G. Hatch Web site, 2001. Available online. URL: http://hatch.senate.gov. Accessed February 10, 2011.

Hickman, Cleveland P. Jr., Larry S. Roberts, Allan Larson, Helen L'Anson, and David J. Eisenhour. *Integrated Principles of Zoology*, 13th Edition. New York: McGraw-Hill, 2006.

Hines, Pamela J. "The Dynamics of Scientific Controversies." *AgBioForum* 4 (2001): 186–193.

Hochedlinger, Konrad. "Your Inner Healers." *Scientific American* 302: (May 2010): 47–53.

Hoggan, Karen. "Neem Tree Patent Revoked." BBC Web site, May 11, 2000. Available online. URL: http://news.bbc.co.uk. Accessed February 10, 2011.

Holmes, Bob. "Altered Animals: Creatures with Bonus Features." *New Scientist* Web site, July 14, 2010. Available online. URL: http://www.newscientist.com. Accessed February 10, 2011.

Houdebine, Louis-Marie. "Transgenic Animal Models in Biomedical Research." *Methods in Molecular Biology* 360 (2007): 163–202.

Huang, David T. and Mehrnaz Hadian. "Bench-to-Bedside Review: Human Subjects Research—Are More Standards Needed?" *Critical Care* 10 (2006): 244. Available online. URL: http://ccforum.com/content/10/6/244. Accessed February 10, 2011.

Human Cloning and Human Dignity: An Ethical Inquiry. Washington, D.C.: The President's Council on Bioethics, July 2002.

Hyun, Insoo. "The Bioethics of Stem Cell Research and Therapy." *The Journal of Clinical Investigation* 120 (2010): 71–75.

"Improving Clinical Use of Stem Cells to Repair Heart Damage." ScienceDaily Web site, July 13, 2010. Available online. URL: http://www.sciencedaily.com. Accessed February 10, 2011.

"In Defense of Evolution." PBS Web site, October 1, 2007. Available online. URL: http://www.pbs.org. Accessed February 10, 2011.

In re Welfare of C.T.L., 722 N.W.2d 484 (Minn. App. 2006).

Johnson, Norman A. *Darwinian Detectives*. New York: Oxford University Press, 2007.

"Joint Statement by Secretaries Clinton and Sebelius on a 1946-1948 Study." U.S. Department of Health and Human Services Web site, October 1, 2010. Available online. URL: http://www.hhs.gov/news. Accessed February 10, 2011.

Jones, Phillip B.C. "Funding of Human Stem Cell Research by the United States." *Electronic Journal of Biotechnology* 3 (April 15, 2000). Available online. URL: http://www.ejbiotechnology.info. Accessed February 10, 2011.

Jones, P.B.C., "Patentability Requirements for Genes and Proteins: Perspectives from the Trilateral Patent Offices." *Journal of BioLaw & Business* 3 (July/August, 2000): 5–15.

Jones, Phill, "Send in the Clones," *Information Systems for Biotechnology News Report*, 1–3 (April 2008).

Jonsen, Albert R. *The Birth of Bioethics*. New York: Oxford University Press, 1998.

Karp, Gerald. *Cell and Molecular Biology*, 6th Edition. New York: John Wiley & Sons, Inc., 2010.

Keim, Brandon. "Parasite-Busting Bugs Throw Fruit Fly Evolution into Overdrive." *Wired Magazine* Web site, July 8, 2010. Available online. URL: http://www.wired.com. Accessed February 10, 2011.

Kelly, Janet. "U.S. Patent Office Issues Certificates to Uphold WARF Stem Cell Patents." WARF Web site, June 26, 2008. Available online. URL: http://www.warf.org. Accessed February 10, 2011.

Kitzmiller v. Dover Area Sch. Dist., 400 F. Supp. 2d 707 (M.D. Pa. 2005).

Kington, Raynard S. "National Institutes of Health Guidelines on Human Stem Cell Research." Stem Cell Information Web site, 2010. Available online. URL: http://stemcells.nih.gov/policy/2009guidelines.htm. Accessed February 10, 2011.

Knight, Andrew. "Chimpanzee Experiments: Questionable Contributions to Biomedical Progress." *AATEX* 14 (2008): 119–124.

Kutz, Gregory. "Direct-To-Consumer Genetic Tests: Misleading Test Results Are Further Complicated by Deceptive Marketing and Other Questionable Practices." Washington, D.C.: U.S. Government Accountability Office, July 22, 2010.

Larson, Edward J. *Summer for the Gods*. Cambridge, Mass.: Harvard University Press, 1997.

Lee, Henry C., Elaine M. Pagliaro, and Katherine Ramsland. *The Real World of a Forensic Scientist*. Amherst, NY: Prometheus Books, 2009.

Leffingwell, Albert. *The Vivisection Question*. New Haven: The Tuttle, Morehouse & Taylor Company, 1901.

Lo, Bernard and Lindsay Parham. "Ethical Issues in Stem Cell Research." *Endocrine Reviews* 30 (2009): 204–213.

Machamer, Peter K., Marcello Pera, Aristeides Baltas, eds. *Scientific Controversies: Philosophical and Historical Perspectives*. New York: Oxford University Press, 2000.

Maguire, A.M., K.A. High, A. Auricchio, J.F. Wright, E.A. Pierce, et al. "Age-dependent Effects of RPE65 Gene Therapy for Leber's Congenital Amaurosis: A Phase 1 Dose-escalation Trial." *Lancet* 374 (2009): 1597–1605.

Maschke, Karen J. "Biobanks: DNA and Research." In *From Birth to Death and Bench to Clinic: The Hastings Center Bioethics Briefing Book for Journalists, Policymakers, and Campaigns*. Mary Crowley, ed. Garrison, NY: The Hastings Center, 2008, 11–14.

Maschke, Karen J. "DNA and Law Enforcement." In *From Birth to Death and Bench to Clinic: The Hastings Center Bioethics Briefing Book for Journalists, Policymakers, and Campaigns*. Mary Crowley, ed. Garrison, NY: The Hastings Center, 2008, 45–50.

Matthews, Qiana L. and David T. Curiel. "Gene Therapy: Human Germline Genetic Modifications—Assessing the Scientific, Socioethical, and Religious Issues." *Southern Medical Journal* 100 (2007): 98–100.

"Mayo's 'Smart' Adult Stem Cells Repair Hearts." Mayo Clinic Web site, August 16, 2010. Available online. URL: http://www.mayoclinic.org. Accessed February 10, 2011.

Melo, Eduardo O., Aurea M.O. Canavessi, Mauricio M. Franco, and Rodolfo Rumpf. "Animal Transgenesis: State of the Art and Applications." *Journal of Applied Genetics* 48 (2007): 47–61.

Moore v. Regents of University of California, 739 P. 2d 479 (Cal. 1990).

Newbury, Umut. "Anti-Frankenfood Movement Grows." *Mother Earth News* (August/September 2004), 18.

Nussbaum, Robert L., Roderick R. McInnes, Huntington F. Willard, and Ada Hamosh. *Thompson & Thompson Genetics in Medicine*, 7th Edition. Philadelphia: Saunders Elsevier, 2007.

Padian, Kevin. "The Evolution of Creationists in the United States: Where Are They Now, and Where Are They Going?" *Comptes Rendus Biologies* 332 (2009): 100–109.

Padian, Kevin and Nicholas Matzke. "Darwin, Dover, 'Intelligent Design' and Textbooks." *Biochemical Journal* 417 (2009): 29–42.

"Patent on Human Embryonic Stem Cells Rejected After Consumer Groups' Appeal." PR Newswire Web site, May 3, 2010. Available online. URL: http://www.prnewswire.com. Accessed February 10, 2011.

Pew Initiative on Food and Biotechnology. "Bugs in the System?" The Pew Charitable Trusts Web site, January 2004. Available online. URL: http://www.pewtrusts.org. Accessed February 10, 2011.

Plomer, Aurora, Kenneth S. Taymor, and Christopher Thomas Scott. "Challenges to Human Embryonic Stem Cell Patents." *Cell Stem Cell* 2 (2008): 13–17.

Pollack, Andrew. "Stem Cell Trial Wins Approval of F.D.A." *The New York Times*, July 30, 2010.

Pollack, Andrew. "Panel Leans in Favor of Engineered Salmon." *The New York Times*, September 20, 2010.

Press, Nancy. "Genetic Testing and Screening." In *From Birth to Death and Bench to Clinic: The Hastings Center Bioethics Briefing Book for Journalists, Policymakers, and Campaigns*. Mary Crowley, ed., 73–78. Garrison, NY: The Hastings Center, 2008.

Prince of Wales. "The Seeds of Disaster." *The Daily Telegraph*, June 7, 1998.

Pritchard, Jonathan K. "How We Are Evolving." *Scientific American* 303 (October 2010): 40–47.

"Questions and Answers About Chimpanzees Used in Research." The Humane Society of the United States Web site, April 9, 2010. Available online. URL: http://www.humanesociety.org. Accessed February 10, 2011.

"Remarks of President Barack Obama—Signing of Stem Cell Executive Order and Scientific Integrity Presidential Memorandum." White House Web site, March 9, 2009. Available online. URL: http://www.whitehouse.gov. Accessed February 10, 2011.

"Report on Findings from the U.S. Public Health Service Sexually Transmitted Disease Inoculation Study of 1946–1948, Based on Review of Archived Papers of John Cutler, MD, at the University of Pittsburgh." U.S. Department of Health and Human Services Web site, 2010. Available online. URL: http://www.hhs.gov/1946inoculationstudy/cdc_rept-std_inoc_study.html. Accessed February 10, 2011.

Ritter, Malcolm. "Adult Stem Cell Research Far Ahead of Embryonic." ABC News Web site, August 2, 2010. Available online. URL: http://abcnews.go.com. Accessed February 10, 2011.

Ruse, Michael, ed. *But Is It Science?* Amherst, New York: Prometheus Books, 1996.

Ruse, Michael and Christopher A. Pynes, eds. *The Stem Cell Controversy: Debating the Issues*, 2nd Edition. Amherst, New York: Prometheus Books, 2006.

Schenker, Joseph G. "Ethical Aspects of Advanced Reproductive Technologies." *Annals of the New York Academy of Sciences* 997 (2003): 11–21.

Scopes, John T. and Presley, James. *The Center of the Storm*. New York: Holt, Rinehart and Winston, 1967.

"Screening Somatic Stem-cells for Safety Shows Promise." *Proceedings of the National Academy of Sciences USA* 107 (2010): 12407.

Segel, Lawrence. "In His Image." *Medical Post (Toronto)* 39 (April 22, 2003): 26.

Shi, Yanhong. "Induced Pluripotent Stem Cells, New Tools for Drug Discovery and New Hope for Stem Cell Therapies." *Current Molecular Pharmacology* 2 (January 2009): 15–18.

Shubin, Neil H. "This Old Body." *Scientific American* 300 (January 2009): 64–67.

Simoncelli, Tania and Sheldon Krimsky. "New Era of DNA Collections: At What Cost to Civil Liberties?" American Constitution Society for Law and Policy Web site, 2007. Available online. URL: http://www.acslaw.org. Accessed February 10, 2011.

Simonelli, Francesca, Albert M. Maguire, Francesco Testa, Eric A. Pierce, Federico Mingozzi, et al. "Gene Therapy for Leber's Congenital Amaurosis is Safe and Effective Through 1.5 Years After Vector Administration." *Molecular Therapy* 18 (2010): 643–650.

Sivakumaran, Theru A. and Marci M. Lesperance. "A PCR-RFLP Assay for the A716T Mutation in the WFS1Gene, a Common Cause of Low-Frequency Sensorineural Hearing Loss." *Genetic Testing* 6 (2002): 229–231.

Slack, Gordy. *The Battle over the Meaning of Everything.* San Francisco: Jossey-Bass, 2007.

Sompayrac, Lauren. *How the Immune System Works*, 3rd Edition Malden, Mass.: Blackwell Publishing, 2008.

State v. Athan, 158 P.3d 27 (Wash. 2007).

Thomson, James A. "Testimony Before the Senate Appropriations Subcommittee on Labor, Health and Human Services, Education Regarding Stem Cell Research." December 2, 1998.

Trounson, Alan. "New Perspectives in Human Stem Cell Therapeutic Research." *BMC Medicine* 7 (2009). Available online. URL: http://www.biomedcentral.com. Accessed February 10, 2011.

"USA: Cultivation of GM plants, 2009." GMO Compass Web site, 2009. Available online. URL: http://www.gmo-compass.org. Accessed February 10, 2011.

Wambaugh, Joseph. *The Blooding.* New York: Bantam Books, 1989.

Washington University v. Catalona, 490 F.3d 667 (8th Cir 2007).

Watt, Fiona M. and Ryan R. Driskell. "The Therapeutic Potential of Stem Cells." *Philosophical Transactions of the Royal Society B* 365 (2010): 155–163.

Weiss, Rick. "Cloning a Previous Hoax?" *The Washington Post*, December 31, 2002.

Weiss, Rick. "U.S. Denies Patent for a Too-Human Hybrid." *The Washington Post*, February 13, 2005.

Wells, D.J. "Gene Doping: The Hype and the Reality." *British Journal of Pharmacology* 154 (2008): 623–631.

Wilmut, Ian. "Cloning for Medicine." *Scientific American* 279 (December 1998): 58–63.

Wood, Daniel B. "'Grim Sleeper' Case Raises Privacy Concerns over Use of DNA." *The Christian Science Monitor* Web site, July 8, 2010. Available online. URL: http://www.csmonitor.com. Accessed February 10, 2011.

Zwaka, Thomas P. "Use of Genetically Modified Stem Cells in Experimental Gene Therapies." In *Regenerative Medicine* 2006, 45–52. Available online. URL: http://stemcells.nih.gov/staticresources/info/scireport/PDFs/Regenerative_Medicine_2006.pdf. Accessed February 10, 2011.

Further Resources

Books

Farndon, John. *From DNA to GM Wheat: Discovering Genetic Modification of Food*. Portsmouth, NH: Heinemann Educational Books, 2008.

Friedman, Lauri S. *Stem Cell Research (Introducing Issues with Opposing Viewpoints)*. Farmington Hills, Mich.: Greenhaven Press, 2009.

George, Linda. *Science on the Edge: Gene Therapy*. Farmington Hills, Mich.: Blackbirch Press, 2003.

Hodge, Russ. *Human Genetics*. New York: Chelsea House, 2010.

Holmes, Thom. *Evolution*. New York: Chelsea House, 2010.

Judson, Karen. *Genetic Engineering: Debating the Benefits and Concerns*. Berkeley Heights, NJ: Enslow Publishers, Inc., 2001.

Nardo, Don. *DNA Evidence*. Farmington Hills, Mich.: Lucent, 2007.

Panno, Joseph. *Gene Therapy*. New York: Chelsea House, 2010.

Simpson, Kathleen. *National Geographic Investigates: Genetics: From DNA to Designer Dogs*. Washington, D.C.: National Geographic Children's Books, 2008.

Web sites

Access Excellence @ the National Health Museum
http://www.accessexcellence.org

The National Health Museum offers general information and images about agricultural biotechnology, DNA fingerprinting, gene therapy, and genetic testing.

Evolution 101
http://evolution.berkeley.edu/evosite/evo101/index.shtml

Created by the University of California Museum of Paleontology, Evolution 101 explains the principles of evolution. The Web site also describes the history of evolutionary thought and current issues in evolution.

Forensic DNA Ethics
http://forensicdnaethics.org/resources/cases

This Web site of the University of Pennsylvania Department of Medical Ethics explores issues in forensic DNA analysis, including forensic phenotyping, the process of gathering clues about appearance from a person's DNA.

Genetic Science Learning Center
http://learn.genetics.utah.edu

The University of Utah provides a wide range of information about DNA, RNA, and proteins. The Web site also offers learning modules about stem cells, cloning, genetically engineered mice, genetic disorders, and gene therapy.

National Human Genome Research Institute
http://www.genome.gov/Issues

The "Issues in Genetics" section of this Web site describes genetic testing, gene patenting issues, genetic discrimination, and explains informed consent.

Stem Cell Information
http://stemcells.nih.gov/info/basics

The National Institutes of Health offer scientific information about stem cells and government policies on stem cell research.

Picture Credits

page:
- 9: © AP Images
- 11: © Bildarchiv Pisarek/akg-images/Newscom
- 13: © Infobase Learning
- 20: © Infobase Learning
- 21: © Infobase Learning
- 23: © Infobase Learning
- 30: © Infobase Learning
- 34: © Infobase Learning
- 35: © Infobase Learning
- 36: © AP Images
- 40: © Infobase Learning
- 42: © Infobase Learning
- 49: © Infobase Learning
- 54: © Infobase Learning
- 55: © Infobase Learning
- 61: © AP Images
- 70: © Infobase Learning
- 71: © Infobase Learning
- 72: © Infobase Learning
- 82: © Infobase Learning
- 84: © Infobase Learning
- 95: © Time & Life Pictures/Getty Images
- 97: © Pictorial Press Ltd/Alamy
- 98: © AP Images

Index

A

abortion, 78
ADA deficiency, 49–50
adenine, 29–30
adenosine deaminase (ADA), 49–50
adult stem cells, 16–19
Agrobacterium tumefaciens, 35–39
Aguillard, Edwards v., 99
Alexander, Leo, 11
Alzheimer's disease, 80
American Anti-Vivisection Society, 8
American Civil Liberties Union, 96
amino acids, 32
amniocentesis, 76
Animal Consultants International, 10
animal experimentation, 7–8, 13, 39–46
animal models, 40
antibodies, 18
antigens, 18
Arkansas, Epperson v., 98, 99
Arteriocyte, 19
Athan (murder suspect), 89
athletes, 54

B

Bacillus thuringiensis, 37
base pairing, 29–31, 72–73
Beagle (HMS), 92
Beck, Charlie, 90
Bernard, Claude, 8–9
biolistics, 36–37
biopiracy, 67
biotechnology. *See* Genetic engineering
blastocysts, 22, 53
Blood Pharming program, 19
body snatchers, 64
bombardment, 36–37
bone marrow stem cells, 17–18, 27
bone marrow transplants, 26
Boyer, Herbert, 33–35
BRCA1, 64

breast cancer, 64
Brennan, William J. Jr., 99
Brinster, Ralph, 39
Bromhall, J. Derek, 53
Bryan, William Jennings, 96, 98–99
Bush, George H.W., 24
Bush, George W., 26
Butler Act, 96–99

C

Canavan disease, 65–66
carrier testing, 75
Center for Bioethics and Human Dignity, 24
Chagas' disease, 41
Chakrabarty, Diamond v., 58, 60–61
Charles (Prince of Wales), 38
chimpanzees, 10
Christian Medical Association, 26
chromosomal disorders, 47, 48
chromosomes, 29, 47, 48, 69, 70
cleavage sites, 32
Clinton, Hillary, 12
Clinton, William J., 24
cloning, 41–43, 44–46, 52–57
CODIS (Combined DNA Index System Program), 84–87, 89–90
codons, 32
Cohen, Stanley, 33–35
Colon, Álvaro, 12
Company of Barber-Surgeons, 64
consent, 11–12, 66, 68
controversies
  about animal experimentation, 7–8, 13, 43–46
  about engineered and cloned animals, 43–46
  about evolution, 95–101
  about forensic DNA analysis, 84–90
  about genetic tests, 77–80

about genetically modified crops, 36, 37–39
about germline genetic modification, 51
about human cloning, 53–57
about human embryonic stem cells, 23–26
about human experimentation, 8–12, 14
about ownership of biological material, 64–68
about patents, 60–64, 66–68
conversion, 65
Cook-Deegan, Robert, 62–64
cord stem cells, 19–20
cosmetics, 78
cows, transgenic, 40–41
creationism, 95–101
crops, genetically modified, 35–39
Cruelty to Animals Act, 8
CSI effect, 86
Cutler, John C., 12
cytosine, 29–30

D

Darrow, Clarence, 97, 98–99
Darwin, Charles, 92–93
databases, 84–90
Defense Advanced Research Projects Agency (DARPA), 19
Department of Energy, 80
designer genes, 54
Devine, Scott E., 74
diagnostics, patents and, 62
Diamond v. Chakrabarty, 58, 60–61
differentiation, overview of, 15–16
direct-to-consumer (DTC) testing, 78–79
diseases, 75–77
DNA (deoxyribonucleic acid). *See also* Genetic engineering
  detection of sequence differences in, 69–75
  genetic testing and, 69–75
  methylation of, 43
  mitochondrial, 45
  overview of, 29
  patents and, 62
DNA Analysis Backlog Elimination Act, 87, 88
DNA fingerprinting, 81–83, 88
DNA polymerases, 71–72
DNA profiling, 83–88
Doctors' Trial, 11
Doll, John, 63

Dolly (sheep), 41
doping, 54
double helix structure, 30, 31
Dover Area School District, Kitzmiller v., 101
Drosophila neotestacea, 100
Drugstore Conspiracy, 96–98

E

EcoRI enzyme, 33
education, evolution and, 95–101
Edwards v. Aguillard, 99
electrophoresis, 71
embryonic stem cell lines, 22
embryonic stem cells
  controversies about, 23–26
  non-embryonic stem cells vs., 26–28
  patent disputes, tissue ownership issues of, 66–68
embryos, debate over, 28
employment rights, 79–80
enzymes, 32–33, 70
epigenetic effects, 43, 51
Epperson v. Arkansas, 98, 99
erythropoietin, 54, 62
eugenics, 9–11, 53–54
European Patent Office, 67
evolution, 91–101
ex vivo gene therapy, 49, 50

F

familial searching, 89–90
felony offenders, 87
fermentation, 58
fertility clinics, 21
First Amendment, 99
FitzRoy, Robert, 92
Food and Drug Administration (FDA), 22–23, 44–46
Foreman, Carol Tucker, 38
forensic science, 45, 81–84
Fortis, Abe, 99
fossils, 92
Fourth Amendment, 88
Frankenfood, 36, 38
Freiden, Thomas R., 12
funding for stem cell research, 24–26

G

Galápagos Islands, 92
gender selection, 78
gene expression, 29–32

Index

gene therapy, 48–51, 54
General Electric Company, 60–61
genes, 31, 75–77
genetic code, 32
genetic engineering
  in animals, 39–41
  controversies about, 43–46
  of crops, 35–39
  human germlines and, 51–52
  in humans, 57
  overview of, 32–35
  patents and, 60–61
Genetic Information Nondiscrimination Act, 79–80
genetic risk factor tests, 76
genetic testing
  controversies about, 77–80
  detection of DNA sequence differences, 69–75
  testing for genes and genetic function related to a disease, 75–77
genetic uniformity, 54
genomic imprinting, 43
Germany, 9–11
germline genetic modification, 51–52
Geron Corporation, 23
Gill, Peter, 83
Golden Rice, 37
Government Accountability Office (GAO), 79
grave robbery, 64
Great Ape Protection Act, 10
Grim Sleeper case, 90
growth factors, 54
guanine, 29–30
Guatemala, 12

H

HaeII enzyme, 73–75
Hatch, Orrin G., 25
hearing loss, 73
heart stem cells, 27
hematopoietic stem cells, 15, 17, 50
hiccups, 94
In His Image: The Cloning of Man (Rorvik), 52–53
hitchhikers, evolutionary, 100
hoaxes, 53
hormones, 44, 54
Human Embryo Research Panel, 24
human embryonic stem cells. *See* Embryonic stem cells

human experimentation, controversy about, 8–12, 14
Human Genome Project, 74, 80
Humane Society, 10
Huntington's disease, 76
hybrids, 63
hygiene hypothesis, 94

I

Iguanas, 92
immune system, 17–18, 41
imprinting, 43
in vitro fertilization, 75
in vivo gene therapy, 49
induced pluripotent stem cells, 52
infertility clinics, 21
informed consent, 11–12, 66, 68
insects, 41
insurance, 79–80
intelligent design, 99–100
Introduction to the Study of Experimental Medicine (Bernard), 8–9

J

Jackson, Joseph, 67
JAK2 gene, 75
Jeffreys, Alec, 69–70, 81–83
Jones, John E. III, 101
jumping genes, 74
Justice for All Act, 87

K

Kaenike, John, 100
Kitzmiller v. Dover Area School District, 101
Knight, Andrew, 10
Kutz, Gregory, 79

L

LDIS. *See* Local DNA Index System
Lee, Henry C., 86
Leffingwell, Albert, 7–8
leukemia, 17, 65
ligase, 33, 34
livestock improvements, 40–41, 44
Local DNA Index System (LDIS), 85
Loring, Jeanne, 68

M

macroevolution, 93
malaria, 94
marrow transplants, 17–18, 27

methylation, 43
mice, 39, 41–42, 52
microevolution, 93
microinjection, 39, 40
microprojectile bombardment, 36–37
Miller, Kenneth, 101
minisatellites, 81, 83
mitochondria, 45
mitochondrial disorders, 47, 48
molecular photocopying, 72
Moore, John, 65
moral convictions, 23–26, 28
mosquitoes, 41
mRNA (messenger RNA), 31
multigene disorders, 47–48
multipotency, 15, 17
mutations, 26

N

National DNA Index System (NDIS), 85
National Institutes of Health (NIH), 24–26, 68, 80
Nazi Germany, 9–11
neem tree oil, 67
newborn screening, 76
Newman, Stuart, 63
Nightlight Christian Adoptions, 26
non-embryonic stem cells, 16–20, 26–28
nuclear transfer, 41, 53
nucleic acids, 29
nucleotides, 29
nucleus, 29
Nuremburg Code, 11
nutrigenomics, 79

O

Obama, Barack, 12, 26, 55, 68
oil, biodegradation of, 60–61
organelles, 45
The Origin of Species (Darwin), 93
ownership of biological material, 64–68

P

paleontology, 99
Palmiter, Richard, 39
Pasteur, Louis, 58
patent infringement, 60
patents, 58–68
Patriot Act, 87
PCR (polymerase chain reaction), 71–73, 83
Peay, Austin, 96

personalized medicine, 76–77, 78–79
pest control, 41
pharmaceuticals, 41
phosphorus pollution, 40–41
pigs, transgenic, 40–41, 59
piracy, 67
Pitchfork, Colin, 82–83
plasmids, 32–34
Plomer, Aurora, 66
pluripotency, 15, 17, 22, 28, 52
polycythemia, 75
polymers, 29
polymorphisms, 70–71
predictive genetic tests, 76, 78–79
pre-implantation diagnosis, 75
prenatal testing, 75–76, 78
Press, Nancy, 78–79
primers, 72
privacy statutes, 80
proliferation, 16
protein synthesis, 31–32

R

Reagan, Ronald, 24
The Real World of a Forensic Scientist (Lee), 86
recombinant DNA technology, 32
red blood cells, 19
rejection, 18, 20
religion, 23–26, 28, 38, 51, 95–101
reproductive cloning, 53–57
restriction enzymes, 32–33, 70, 73–75, 81
restriction fragment length polymorphisms (RFLP), 70–71, 83
Resurrection Men, 64
Reverby, Susan, 12
RFLP. *See* Restriction fragment length polymorphisms
rice, 37
RNA (ribonucleic acid), 31
Rorvik, David, 52–53
roundworms, 100
rubber trees, 67

S

safety, 50–51, 52. *See also* Controversies
salmon, transgenic, 44
Sarbah, Andrew, 81–82
Schneider, Michael, 27
scientific research, patents and, 62
Scopes Monkey Trial, 96–99
Scott, Christopher Thomas, 66

Index   123

SDIS. *See* State DNA Index System
Sebelius, Kathleen, 12
selective breeding, 9–10, 44
self-renewal, 16
sex offenders, 87
sheep, transgenic, 41
short tandem repeats (STR), 83–84
Shubin, Neil, 94
sickle cell disease, 94
single-gene disorders, 47
somatic cells, 51
space program, 9
spinal cord injuries, 23
Spiroplasma spp., 100
State DNA Index System (SDIS), 85
Stearns, Stephen, 94
stem cells, 16, 52–54. *See also* Embryonic stem cells; Non-embryonic stem cells
STR. *See* Short tandem repeats
subculturing, 22
sugar-phosphate backbone, 29–30
survival of the fittest, 94
Sweet, Robert W., 64
syphilis, 12

T

tandem repeats, 82
Taymor, Kenneth S., 66
teaching of evolution, 95–101
television shows, 86
terminal differentiation, 15
Terzic, Andre, 27
therapeutic cloning, 53, 54
therapeutic proteins, 62, 64
The Thief at the End of the World (Jackson), 67
Thomson, James A., 20–22, 24, 25, 66–68
thymine, 29–30
tissue replacement, 22
tissue samples, ownership of, 64–66

totipotency, 15
transcription, 31–32, 43
transgenes, 32
translation, 31–32
transplants, 17–18, 22, 26, 41
transposons, 74

U

umbilical cord stem cells, 19–20
U.S. Patent and Trademark Office (USPTO), 59, 63, 66–68
U.S.A. Patriot Act, 87
utility patents, 59

V

vectors, 32–34, 36, 39, 48–49
viruses, 39, 48–49, 52
vivisection, 7–8, 13, 39–46
The Vivisection Question (Leffingwell), 7–8

W

Wallace, Alfred Russel, 93
warfarin, 77
warrants, 88
Waters, Acacia, 9
Watson, James, 62
white blood cells, 17–18
Wickham, Henry, 67
Wolfram syndrome (WFS1) gene, 73–75
World Anti-Doping Agency, 54
World War II, 9–11

X

xenotransplantation, 41

Y

Yamanaka, Shinya, 52

Z

zygotes, 15

About the Author

Phill Jones earned a Ph.D. in physiology/pharmacology from the University of California, San Diego. After completing postdoctoral training at Stanford University School of Medicine, he joined the Department of Biochemistry at the University of Kentucky Medical Center as an assistant professor. Here, he taught topics in molecular biology and medicine and researched aspects of gene expression. He later earned a JD at the University of Kentucky College of Law and worked for 10 years as a patent attorney, specializing in biological, chemical, and medical inventions. Dr. Jones is now a full-time writer. His articles have appeared in *Today's Science on File*, *The World Almanac and Book of Facts*, *History Magazine*, *Forensic Magazine*, *Genomics and Proteomics Magazine*, *Encyclopedia of Forensic Science*, *The Science of Michael Crichton*, *Forensic Nurse Magazine*, *Nature Biotechnology*, *Information Systems for Biotechnology News Report*, *Law and Order Magazine*, *PharmaTechnology Magazine*, and educational testing publications. His books, *Sickle Cell Disease* (2008), *The Genetic Code* (2010), and *Kingdoms of Life* (2011) were published by Chelsea House. For more than five years, he has taught an online course in forensic science for writers. His science fiction/mystery novella, *Thin Ice* (2010), and numerous short stories have also seen publication.